T0184750

Re-thinking Mobility Poverty

This book seeks to better conceptualise and define mobility poverty, addressing both its geographies and socio-economic landscapes. It moves beyond the analysis of 'transport poverty' and innovatively explores mobility inequalities and social construction of mobility disadvantages.

The debate on mobility poverty is gaining momentum due to its role in triggering social exclusion and economic deprivation. In this light, this book examines the social construction of mobility poverty by delving into mobility patterns and needs as they are differently experienced by social groups in different geographical situations. It considers factors such as the role of transport regimes and their social value when analysing the social construction of individual's mobility needs. Furthermore, the gaps between articulated and unarticulated needs are identified by observing actual travel patterns of individuals. The book offers a comparison of the global phenomenon through fieldwork conducted in six different European countries – Greece, Portugal, Italy, Luxembourg, Romania and Germany.

This book will be useful reading for planners, sociologists, geographers, mobility/transport researchers, mobility advocates, policy-makers and transport practitioners.

Tobias Kuttler is a geographer and urban planner, working on urban development, mobility and social disadvantage. He is a researcher at TU Berlin and PhD Candidate at TU Munich. He is writing his PhD dissertation on the transformation of the taxi industry in Mumbai, India.

Massimo Moraglio is a Senior Researcher at Technische Universität Berlin and Coordinator of the MBA "Sustainable Mobility". His work explores the relationship between technology and society, focusing on scenarios and transitions. He is a member of the editorial board of *Mobilities* and *Applied Mobilities*, and editor-in-chief of *The Journal of Transport History*.

Transport and Society
Series Editor: **John D. Nelson**

This series focuses on the impact of transport planning policy and implementation on the wider society and on the participation of the users. It discusses issues such as: gender and public transport, travel for the elderly and disabled, transport boycotts and the civil rights movement etc. Interdisciplinary in scope, linking transport studies with sociology, social welfare, cultural studies and psychology.

The Mobilities Paradigm
Discourses and Ideologies
Edited by Marcel Endres, Katharina Manderscheid and Christophe Mincke

Ports as Capitalist Spaces
A Critical Analysis of Devolution and Development
Gordon Wilmsmeier and Jason Monios

Non-motorized transport integration into urban transport planning in Africa
Edited by Winnie Mitullah, Marianne Vanderschuren and Meleckidzedeck Khayesi

Assembling Bus Rapid Transit in the Global South
Translating Global Models, Materialising Infrastructure Politics
Malve Jacobsen

Re-thinking Mobility Poverty
Understanding Users' Geographies, Backgrounds and Aptitudes
Edited by Tobias Kuttler and Massimo Moraglio

For more information about this series, please visit: www.routledge.com/Transport-and-Society/book-series/TSOC

Re-thinking Mobility Poverty

Understanding Users' Geographies, Backgrounds and Aptitudes

Edited by
Tobias Kuttler and
Massimo Moraglio

Routledge
Taylor & Francis Group

LONDON AND NEW YORK

First published 2021
by Routledge
2 Park Square, Milton Park, Abingdon, Oxon OX14 4RN

and by Routledge
52 Vanderbilt Avenue, New York, NY 10017

Routledge is an imprint of the Taylor & Francis Group, an informa business

British Library Cataloguing-in-Publication Data
A catalogue record for this book is available from the British Library

Library of Congress Cataloging-in-Publication Data
A catalog record has been requested for this book

ISBN: 978-0-367-33330-0 (hbk)
ISBN: 978-0-367-33331-7 (ebk)

Typeset in Times New Roman
by codeMantra

I dedicate this work/this volume to my parents, Erika and Franz (TK)

Contents

Figures

Tables

Contributor biographies

Stefano Borgato (MSc in Environmental and Land Planning Engineering, MA in Urban Studies) has more than five years' experience in transport planning, research in innovative modes of transportation and sustainability-related projects. Since 2018, he has been working as a Transport Planner and Policy Analyst for TRT Trasporti e Territorio in Milan.

Simone Bosetti (MSc in Civil Engineering – Transport) has 20 years' experience in transport planning at international, regional and local levels. On behalf of TRT Trasporti e Territorio, he is the Coordinator of the HiReach project, funded under the EU Horizon 2020 research programme, concerning transport equity and inclusion.

Cosimo Chiffi (MSc in Maritime and Transport Economics) has 15 years' experience in transport policy analysis and economics. His expertise lies in transport demand analysis, mobility planning, environmental and socio-economic assessment of transport projects and policies. Since 2005, he has been working as a Consultant for TRT Trasporti e Territorio.

Patrick van Egmond has an MSc in Business and Transport Economics from the Free University of Amsterdam (2001). Patrick is the CEO of LuxMobility, a Luxembourgish based consultancy. He manages several teams in Europe and Africa in the fields of transport, logistics, rural and urban mobility and smart cities.

André Freitas graduated in Sociology (2006) with post-graduate qualifications in Planning and Human Resources. Since 2018, he has been working as a Consultant for TIS, most notably on the HiReach project, working with vulnerable social groups and investigating the feasibility of new digital solutions tailored to their needs.

Razvan Andrei Gheorghiu is a Senior Lecturer and Researcher at the Department of Electronics in Transport of the "Politehnica" University of Bucharest. With over 15 years' experience, his areas of interest include

Intelligent Transport Systems and Computer Science. He is the author of over 50 academic articles and five books.

Delphine Grandsart is a Senior Researcher for the European Passengers' Federation. She has worked on several research projects including USEmobility, IT2Rail, GOF4R, CIPTEC and HiReach, where she was co-author of reports on diverse topics such as transport poverty, user groups' needs, passenger rights, market uptake of new mobility services and technologies.

Valentin Iordache is a Senior Lecturer and Researcher at the Department of Electronics in Transport of the "Politehnica" University of Bucharest. With over 17 years' experience, his areas of interest include intelligent transport systems, sensors and data acquisition and computer networks. He is the author of over 30 academic articles and three books.

Akrivi Vivian Kiousi is a Senior Innovation Manager at INTRASOFT International S.A. and Head of the Transport Lab in its Research and Innovation Development Department. She has 20 years' experience in the field of big data, transport, telecommunications, e-government, human computer interaction, internet based technologies, community portals, promotion and dissemination strategies.

Mariza Konidi is a Communication and Project Management Specialist in the Research and Innovation Department of INTRASOFT International. She has been involved in implementing and leading dissemination-related activities in research projects. Mariza brings more than ten years' experience, with a strong background in marketing, communications and project management.

Tobias Kuttler is a geographer and urban planner, researching urban development, mobility and social disadvantage. He is a Research Associate at Technische Universität Berlin and PhD candidate at the Technical University of Munich. He is currently writing his PhD dissertation on the transformation of the taxi industry in Mumbai, India.

Silvia Maffii (PhD in Urban and Regional Planning) has extensive professional and research experience in transport economics and planning. Since 1992, she has been the co-founder and Managing Director of TRT Trasporti e Territorio. Her main expertise lies in open issues in transport policies, cost-benefit analysis and socio-economic impacts assessment.

Patrizia Malgieri (PhD in Urban and Regional Planning) has more than 25 years' experience in transport and mobility planning. She has extensive knowledge of urban and regional transport plans, mobility projects and research activities mainly in the fields of sustainable and gender mobility. She is the Director of the urban transport planning division at TRT Trasporti e Territorio.

Massimo Moraglio (PhD) is a Senior Researcher at Technische Universität Berlin and Coordinator of the MBA "Sustainable Mobility". His
work explores the relationship between technology and society, focusing
on scenarios and transitions. He is a member of the editorial board of
Mobilities and *Applied Mobilities*, and editor-in-chief of *The Journal of
Transport History*.

Vasco Reis is an experienced researcher and consultant. Combining practical industry experience with deep academic research, he provides innovative insights about the present and future of mobility and transportation
systems. His experience includes the participation in and management
of more than 30 projects and authorship of over 70 peer-reviewed
publications.

Dariya Rublova is involved in European and Greek research projects focused on regional development and technological change, innovation
policy and development, transport and sustainable mobility and web entrepreneurship. She studied Economic Sciences and completed her MSc
in Applied Economics and Finance at the National and Kapodistrian
University of Athens.

Mimi Sheller, PhD, is Professor of Sociology and founding Director of the
Center for Mobilities Research and Policy at Drexel University in Philadelphia. She is founding co-editor of the journal *Mobilities* and past President of the *International Association for the History of Transport, Traffic
and Mobility*. She is the author of several books on mobility.

Joanne Wirtz has an MA in Geography and Spatial Planning, with a specialisation in geographical modelling, from the University of Luxembourg
and a BSc in Geography, with a specialisation in urban and regional
planning, from the University of Bonn. She currently works as a Mobility
and Transport Consultant and Principal Researcher for LuxMobility.

Acknowledgements

Apart from the authors of this volume, the following persons contributed to this publication project by performing fieldwork in the respective countries: Norman Döge, Valentina Fava, Mahendra Singh Chouhan, Johanna Helene Ostendorf (Technische Universität Berlin, Germany), Andrea Selan (TRT, Italy), Martin Kracheel, François Sprumont (LUXM, Luxembourg), Maggie Poupli (INTRA, Greece), Susana Castelo, Fátima Santos (TIS, Portugal). The book editors and authors express their gratitude to all the contributors.

The book editors and authors are highly indebted to the fieldwork participants, especially the focus group participants, who generously opened up to the research teams and provided first hand experiences and information. The editors and authors would like to thank the following institutions and their representatives:

- **Germany:** Municipality of Wendlingen, Citizen bus umbrella organisation "proBürgerbus" Baden-Württemberg, Kreisseniorenrat Esslingen (County seniors' council), volunteer ridesharing initiative "SuseMobil" Filderstadt, Municipality of Aichwald, Bürgerbus Aichwald (citizen bus initiative), Municipality of Frickenhausen, Arbeiterwohlfahrt (Workers' Welfare Association) Esslingen, Social service for refugees of Esslingen county, Voluntary mobility initiative for refugees "Fahrradfüchse" (bicycle foxes) Donzdorf, Municipality of Donzdorf, Ministry for Transport Baden-Württemberg – Division for digitisation and mobility in rural areas, Transport Department of County of Esslingen, Transport Authority Metropolitan Region Stuttgart VVS
- **Greece:** Gymnasium of Chalkeio, Technical and Professional Lyceum, Filoti – Naxos, 1st Primary School of Naxos, Taxi company owner/President of the Skadou Village Community, Local Bus Naxos, Municipality of Naxos and Small Cyclades, Express Skopelitis ferry boat – Small Cyclades Line, Centre of Creative Work for People with Disabilities
- **Italy:** Coop Terrarossa, Fondazione Le Costantine, Rete SanFra, Comune di Patù, Consigliera Nazionale di Parità, Calzaturificio Sud Salento, Università di Lecce, Ferrovie Sud-Est SpA, ASL Lecce – Comitato Unico di Garanzia

- **Luxembourg**: Ministry of transport, Emile Weber/WebTaxi, Kussbuss, LISER, Forum pour l'emploi, Bummelbus (service of FPE)
- **Portugal:** Municipality of Torres Vedras, Parish of Maxial and parish of Carmões, Associação de Socorros da freguesia de Dois Portos, Municipality of Guarda, Parish of Jarmelo, Valhelhas and Videmonte, Parish of "Runa e Dois Portos", Barraqueiro
- **Romania:** Municipality of Buzau, Trans Bus, Social Democratic Party's Elderly Association, Vox Civica Association, Arts High School, "Margareta Sterian", Vision Impaired People Association from Romania

The book editors and authors furthermore would like to thank the local organisations and their representatives, who worked with the research teams locally and thereby made the fieldwork possible:

- **Germany:** Dr Martin Schiefelbusch (Nahverkehrsgesellschaft Baden-Württemberg NVBW), Fred Schuster (Municipality of Wendlingen, Citizen bus umbrella organisation "proBürgerbus" Baden-Württemberg) Karl Praxl and Tino Marling (Kreisseniorenrat Esslingen (County seniors' council), volunteer ridesharing initiative "SuseMobil" Filderstadt and Altenzentrenförderverein Filderstadt e.v.), Albert Kamm (Municipality of Aichwald and Bürgerbus Aichwald (citizen bus initiative)), Regine Theimer (Municipality of Frickenhausen), Sybille Hegele, Lilith Cattaneo, Farhad Hage and team (Arbeiterwohlfahrt (Workers' Welfare Association) Esslingen/Neckar), Reinhold Sawatzki (voluntary mobility initiative for refugees "Fahrradfüchse" (bicycle foxes) Donzdorf).
- **Greece:** Municipality of Naxos and Small Cyclades, Irakleia Municipal Unit, Centre of Creative Work for People with Disabilities
- **Italy:** Cirino Carluccio (Patù Municipality – Città Fertile, Tricase Municipality – Città Fertile)
- **Luxembourg**: Ministry of Transport, Emile Weber/WebTaxi, Kussbuss, LISER, Forum pour l'emploi, Bummelbus (service of FPE)
- **Portugal:** Horácio Brás (Municipality of Guarda)
- **Romania:** Alexandru Buscu (Social Democratic Party's Elderly Association), Calin Iosif (Vox Civica Association), Veronica Tudose (Arts High School, Margareta Sterian), Iulian Caraman (Vision Impaired People Association from Romania)

We owe gratitude to Alison Hindley Chatterjee for her patient and thorough proof-reading and editing of the draft manuscript. The editors would like to thank Theresa Zimmermann for her helpful and poignant comments on the draft version of the book.

Disclaimers

This document is part of HiReach, a project that has received funding from the European Union's Horizon 2020 research and innovation programme under grant agreement No. 769819.

The sole responsibility for the content of this document lies with the authors. It does not necessarily reflect the views of the European Commission.

The authors acknowledge support by the Open Access Publication Fund of Technische Universität Berlin.

Preface

Silvia Maffii and Simone Bosetti

This book has been prepared by an international team of multidisciplinary experts and researchers sharing a common interest in the themes of the mobility of vulnerable groups, accessibility of peripheral and rural areas, "transport poverty" and a common belief that the issues of inclusive mobility and equity are becoming more and more relevant due to societal changes and new transport policy paradigms of "mobility for all".

The expert team has been cooperating in the three-year (2017–2020) research project HiReach: "High reach innovative mobility solutions to cope with transport poverty". The project was funded under the "Smart, green and integrated transport" thematic area of the EU Framework Programme for Research and Innovation Horizon 2020.

The topic addressed by the HiReach project consists in improving accessibility, inclusive mobility and equity by obtaining a comprehensive view of the mobility needs and capabilities of vulnerable social groups living in unprivileged areas and by developing new tools and business models to address the needs of these people and to promote inclusive and participative mobility.

The HiReach project partnership bridges diverse sets of specialised skills and builds on strong integration and high complementarity of expertise. The partners include two universities (Technical University of Berlin and Politechnica University of Bucharest) to expand the competences in the academic field; one international organisation (the European Passengers' Federation – EPF) to enrich the perspective of public transport passengers across Europe; four consultancies in the field of transport planning, economics and innovation (Intrasoft, LuxMobility, TIS and TRT Trasporti e Territorio, the project leader); and one partner with strong links to the European start-ups ecosystem, Productised. Covering six countries – Germany, Greece, Italy, Portugal, Luxembourg and Romania – the partnership guarantees a wide geographical coverage.

Seven vulnerable social groups have been specifically targeted in the project: low income and unemployed people; elderly people; children and young people; people with reduced mobility; migrants and ethnic minorities; women; and people living in rural and deprived areas. In addition, six study regions have been selected to cover all these typologies: the districts of Esslingen and Göppingen (Germany); Naxos and Small Cyclades (Greece); Southern Salento, inner area (Italy); Guarda and Torres Vedras (Portugal); North and South-East Luxembourg (Luxembourg); and Buzau (Romania).

The content of this book builds on the work carried out in the analytical phase of the project. This phase was dedicated to an in-depth comprehensive review of the mobility needs and capabilities of the seven vulnerable social groups living in different contexts across Europe. The subsequent phases were devoted to the exploration and development of innovative inclusive mobility solutions.

The mobility needs of vulnerable users

The focus of the analytical phase was the exploration of the travel behaviour and social habits of the targeted vulnerable groups as well as the assessment of their travel demand and mobility needs. Importantly, despite growing awareness, transport poverty has not been comprehensively described as a concept yet. Academia, policy-makers and practitioners still need to define and understand the full implications of this phenomenon. Consequently, the first step required the elaboration of the concept of transport poverty including a complex assessment of inequality and disadvantage, distinguishing between transportation-related disadvantage, social disadvantage and social exclusion.

Nobody would deny that the availability of transport options is vital to reach essential opportunities such as jobs, education, shops and friends. It is recognised that mobility is key to full participation in society and meaningful living. However, how this differs according to various economic, social and cognitive parameters and to what extent transport-related deficiencies affect individuals and groups are issues still to be fully explored. More specifically, two open questions are:

- how do the socio-economic and socio-demographic positions of individuals, their skills, personal attitudes, perceptions and aspirations affect transport poverty? and
- how does the surrounding environment where they live – i.e. the urban, rural or peri-urban setting – affect both an individual's transport-related difficulties and social position?

The HiReach project tried to adopt this complex local perspective, identifying common elements that produce transport poverty across Europe to help outline joint solutions in order to eliminate it.

Transport poverty has been investigated by looking at the most vulnerable groups: people experiencing material deprivation or physical impairment; migrants or those with an ethnic minority background; and different socio-demographic characteristics (age, gender) across different European urban, peri-urban and rural areas. It is important to underline that each vulnerable group exhibits unique mobility features and that people could be permanently or temporarily part of a vulnerable group during their lifetime and/or could belong to more than one group. In order to account for the complexity of this combination of social, contextual and relational disadvantages, each project's study region has been associated with two or more transport vulnerable social groups. The social groups taken into consideration include:

- *Low income and unemployed* people, whose mobility is significantly limited by transport costs and who are particularly reliant on local public transport services. If they live *in rural and deprived areas* poorly served by public transport services, the most common way to fulfil mobility needs is to own a car, which leads to the well-known phenomenon of forced car ownership.
- *Elderly people,* who endure different forms of transport disadvantage due to diminishing physical and cognitive capabilities that, combined with progressive digitalisation, could make their access to public transport services challenging.
- *Children and young people,* who show mobility limitations due to their lack of autonomy or responsibility which makes them reliant on their parents to fulfil their mobility needs.
- *Migrants and ethnic minorities,* whose low income, language and cultural barriers reduce their ability to purchase services or means of transport.
- *Women,* whose transport disadvantages are related to their generally more complex travel patterns being engaged in childcare, domestic work and caring for the elderly, and to safety issues, which are not properly addressed by traditional commuter services.
- *People with reduced mobility,* who require appropriate and specific attention as well as an adaptation of the transport services made available to all other passengers and their particular needs.
- *People living in rural and deprived areas,* who are transport-vulnerable as they are usually poorly served by public transport services: low frequencies, limited connectivity, long transit times, long distances to cover and less proximity to the transit stations.

The next steps: looking at the solutions

The analytical work highlighted that to respond to the mobility needs of the vulnerable user group, the solutions should have a public transport component and either improve or complement the mass public transport system. The assessment of the current transport offer has identified three main clusters of mobility options and services as fields of application for inclusive mobility:

– *Publicly contracted transport services*: school buses, door-to-door minibus/van services for people with disabilities or healthcare needs, but also demand responsive transport services in low density and rural areas or at off-peak times, subsidised taxi companies, own-account services of pupils or people with temporary or permanent disabilities. Adaptations and overall improvement of conventional public transport (typically scheduled bus and rail services) can make these services better "fitted for all".
– *More traditional market-based mobility services*: on-street taxis and pre-booked private hire vehicles as well as other solutions that are nowadays offered by companies that can directly or indirectly operate a fleet. New business models have been developed within the so-called shared economy paradigm, facilitated by technology advancements.
– *Community-based mobility options* include community transport services provided by non-profit entities receiving minimum subsidies (e.g. in rural areas or for special transport services), informal or peer-to-peer ride-sharing (car-pooling), shared "village cars" or peer-to-peer car sharing and community-owned bus services often referred to as "citizen buses".

The final objective of HiReach is to find and exploit new business ideas that reach low-accessibility social groups and areas. Following the analytical stage, an exploratory phase of the project has critically assessed the limits of the current transport offer and has identified innovative frameworks and business models of new inclusive, affordable and reliable transport solutions. The purpose of these options is to fit the capabilities and attitudes of different targeted vulnerable groups, who could also be co-users and co-owners, and to provide users with suitable, reliable and affordable transport options to reach nearby or distant destinations in reasonable travel time and by using vehicles/infrastructures/services that are fully accessible, safe and relatively easy to use.

Among the innumerable examples, the HiReach project has handpicked 20 transport solutions to understand their accomplishments and limitations, to analyse their business models and management scheme and to examine the social/technological innovations brought by each solution. Those solutions with the highest potential for replication in other regions/countries have been studied in depth to understand the reason behind their success and the team looked in particular at (i) new organisational and business models; (ii) services that help to upgrade the present transport offer's image and attractiveness; (iii) enabling ICT solutions and interoperability rules to increase access and usage; and (iv) new forms of transport services based on the sharing economy and community-based principles.

This exploratory step has resulted in the elaboration of a set of recommendations on how to implement inclusive, affordable and reliable mobility solutions, which can be summarised as follows:

- for users and communities to consider themselves owners, producers and consumers of inclusive mobility services and be the first active component of policies that do not necessarily entirely rely on public interventions;
- for entrepreneurs and investors to develop services that are modular, likely to be upscaled and that address different target groups and to be open not only to accommodate requests from regular passengers but also to address other emergent groups' needs; and finally
- for policy-makers to investigate transport poverty, duly consider community engagement and properly secure permanent funding for tackling transport poverty issues.

Recommendations, particularly for users and investors, were the starting point for the design and launch of the final phase of the HiReach project, that of development. This phase consisted in the development and testing of new mobility solutions tailored to the vulnerable groups through the creative work of start-ups, innovative entrepreneurs and local communities.

This is what happened within the HiReach Startup Lab, an acceleration program for innovative transport solutions that address transport poverty in Europe. In practical terms, several start-ups and young entrepreneurs were offered a unique opportunity to prototype inclusive mobility solutions for vulnerable groups with the support of a dedicated host company, professional business coaching, showcasing opportunities and financial support.

Foreword

A mobility justice lens on mobility poverty

Mimi Sheller

This book offers an important new framework for thinking about "mobility poverty". It extends beyond existing ideas of transport poverty by utilising new concepts such as social capital, motility, transport regimes and the social construction of transport needs. Drawing on case studies from several European countries, it encompasses different societal dimensions of mobility poverty. These include gender, income and employment, age, reduced mobility and the experiences of migrants, ethnic minorities, children and young people. Beyond that though, it offers important lessons for transportation planning around the world, by pushing us to connect mobility poverty to wider issues of mobility justice. It is this concern that I want to draw attention to here.

As movements and policy-makers around the world seek to create more sustainable mobilities, it is crucial to pay attention to the intersectionality of various kinds of mobility injustice and the impact of changing infrastructural and regulatory systems on those affected by "mobility poverty". Without a clear conceptualisation of *mobility justice,* energy transition policies will continue to reproduce patterns of land use and investment that end up benefitting mainly the kinetic elite by leaving in place the advantages for those who already exercise mobility power. Why should the majority be subjected to mobility austerity (carbon pricing, congestion pricing, higher fuel taxes) or to greater policing of their movements (new border regimes, security surveillance, policing of travellers), when it is the small minority of high consumers of mobility and energy who have produced the vast majority of harmful emissions that contribute to climate change and also drive climate-related migration?

This is the poignant question raised by movements of the "marginalised", ranging from the *gilets jaunes,* who have protested fuel taxes in France, to the black, indigenous and people of colour social movements in the United States, such as People for Mobility Justice and Equiticity, who have questioned transport policies that serve the urban elite (Sheller 2018). Policies for carbon pricing, congestion charging or the creation of "safer" bicycling and pedestrian infrastructure should be carefully coupled with affordable housing policies around transit-oriented development to prevent the suburbanisation of poverty into car-dependent exurbs. Simply inserting carbon

pricing, promoting transit-oriented development or subsidising electric vehicles may actually resist transformative change if it leaves unchallenged the
underlying culture of private, individualised automobility and the spatial
and social relations that go along with automobility, including the cultural
discourses that equate personal private mobility with freedom and dominance (Freudendal-Pedersen 2009). Roads and highways dominate the built
landscape and the over-arching mobility culture remains one in which automobility is normalised as freedom and associated with wealth and privilege.

The theory of mobility justice allows us to strengthen and foreground a
vision for transport equity that includes an understanding of the historically
uneven impacts of infrastructure on land use patterns that have created contemporary splintered urbanism, racial and class segregation, lack of accessibility and automobile dependence for those who can no longer afford the
right to the city. Projects such as the "Green New Deal" in various countries
must address the needs of the mobility poor who may not benefit from new
infrastructure projects and may in fact be harmed and displaced by them.
As John Urry showed us, the *dominant system of automobility* is not just a
means of transportation, but is an interlocking system of social practices,
industrial and business networks and ongoing relations between mobility,
energy production, consumption and land use, intertwined in a complex
system that is heavily influenced by powerful mobility regimes that work in
the interest of "kinetic elites" (Sheller and Urry 2000; Urry 2007).

A holistic theory of mobility justice can help us perceive the connections
across different regimes of mobility, spanning the scale of the unsafe mobilities experienced by women, children, the elderly, ethnic minorities and
immigrants in many cities to the displacement of minority and working
class communities by "green gentrification". At the same time, it can expose
the relation between past histories of colonialism and the ease of travel for
global elites occurring alongside the world-wide building of walls, detention
of migrants and the breakdown of international agreements for the reception of refugees. While sustainable transportation is a crucial element of urban spatial justice, the mobility justice perspective can help us see the wider
kinopolitical struggles embedded in these policies.

We need a vision for transport equity that includes an understanding of
the uneven impacts of infrastructure on inequitable land use patterns understood in terms of segregation, lack of accessibility, "splintered urbanism" and "low-carbon gentrification". While building improved, protected
cycling infrastructure has been heavily promoted by cycling advocates; in
many cases such projects have inadvertently produced spaces of inequality
and exclusion (Golub et al. 2019; Cox and Koglin 2020). Recent Vision Zero
policies that increase the enforcement of violations, for example, have in
the United States been found especially detrimental to utility cyclists and
delivery riders, many of whom are racial minorities or immigrants (Lee
et al. 2016). According to Stephen Zavestoski and Julian Agyeman, moreover, there is a "mobility bias" rooted in the neoliberal foundations of the
Complete Streets concept, which has failed "to give voice to the historically

marginalised" and, more broadly, "to approach streets as dynamic, fluid and public social places" (Zavestoski and Agyeman 2014).

There is a growing splintering of mobility systems into hypermobile, friction-free, rich arrays of mobility choices for the kinetic elite, who can afford to live in well-connected cities, versus the displacement, slow modes and burdensome travel for the mobility poor, who can no longer afford to live in places with good access to public transit. Sustainability transitions must also find ways to undo the land use patterns that have contributed to urban sprawl, urban gentrification, unaffordable housing and high automobile dependence amongst disinvested communities, who have been excluded from property-based wealth. We must advocate for defending rights to mobility and accessibility for under-resourced communities (including rural areas), expanding investment in public transit, including bus rapid transit, and improving inter-regional rail services, while also protecting and building affordable housing.

This calls for explicit reparations for the current harmful impacts of uneven infrastructure and uneven mobilities. We need to change land use rules, housing policies and energy grids and connect these projects to advancing mobility justice. This means better planning around the inter-relation of uneven geographies, uneven mobilities and the reproduction of class and racialised space, which is also of course gendered, sexed and abling/disabling. The freedom to move and to remain somewhere, to dwell in a place and to determine one's own personal and familial movements are fundamental elements of mobility justice. Access to sustainable low-carbon systems of transportation and public transit is one fundamental dimension of this struggle, but it is not the only one.

In addition, we must be willing to check the excessive mobility of "kinetic elites" by placing limits on unbridled energy consumption, excess travel and binge flying. Elite mobilities are often hidden or secluded from the public gaze; yet attention to privileged forms of mobility can help us better understand the power inherent in systems of uneven mobility. By recognising how power is exercised through the control of im/mobilities and spatial moorings, we can better understand the experience of mobility poverty. Anthony Elliott, for example, describes the mobile lives of "globals" who practise lifestyles of "detached engagement; floating; speed; networked possibilities; distance from locality and mapping of escape routes" (Elliott 2013). What of the lived experience of those without high motility? Uneven mobilities – including infrastructures of differential mobility and the hoarding of "mobility capital" by the kinetic elite – have produced the "mobility poor" and those who suffer most the public health burdens of the polluting fossil-fuel infrastructures and the sprawling suburbanisation associated with dominant systems of automobility and aeromobility.

The valorisation of friction-free mobility comes at the expense of those whose class, race, gender, age or abilities subjects them to slower travel, insecurity, policing, unsafe travel conditions, lack of access to travel documents and unhealthy low-waged work in which their time and movements are monopolised and controlled. We must be willing to set limits on the dominant systems of private automobility (and aeromobility) not only to reduce the excessive consumption of fossil-fuelled mobility by the kinetic elites but also to repair and prevent splintered urbanism, differential mobilities and the global resource extraction that leads to "climate colonialism". A mobility justice approach incorporates not only transportation justice but also recognises the wider coloniality of mobility regimes and the need for reparations. Intersectional struggles for mobility justice must address legacies of colonialism, present forms of coloniality and confront questions of access, control, ownership and capitalism.

Finally, we must also recognise the importance of democratising transportation planning processes, decision making and evaluation by including community-based organisations, many of which are already mobilising around mobility justice. For without mobility justice we will not achieve planetary sustainability. The path to sustainable mobilities will undoubtedly depend on embodied kinopolitical struggles that directly challenge the dominant system of automobility, calling out the history of green gentrification, the excesses of kinetic elites and the problems of extractive resource colonialism. Bringing many voices to the conversation and opening up transport planning to diverse stakeholders are crucial starting points through which we can attend to the critical analysis of mobility injustice and mobility poverty. I hope this valuable edited collection will be a good starting point to begin those difficult conversations.

References

Cox, Peter and Till Koglin (eds.). 2020. *The Politics of Cycling Infrastructure: Spaces and Inequality.* Bristol: Bristol University Press.

Elliott, Anthony. 2013. "Elsewhere: Tracking the Mobile Lives of Globals." In *Elite Mobilities*, edited by Birtchnell, Thomas and Javier Caletrio, 21–39. London and New York: Routledge.

Freudendal-Pedersen, Malene. 2009. *Mobility in Daily Life: Between Freedom and Unfreedom.* London and New York: Routledge.

Golub, Aaron, Melody Hoffmann, Adonia Lugo and Gerardo Sandoval (eds.). 2019. *Bicycle Justice and Urban Transformation: Biking for All.* London: Routledge

Lee, Do et al. 2016. "Delivering (in)justice: Food Delivery Cyclists in New York City." In *Bicycle Justice and Urban Transformation: Biking for All?* edited by Aaron Golub, Melody Hoffmann, Adonia Lugo and Gerardo Sandoval, 114–129. London: Routledge.

Sheller, Mimi. 2018. *Mobility Justice: The Politics of Movement in an Age of Extremes.* London and New York: Verso.

Sheller, Mimi and John Urry. 2000. "The City and the Car." In *International Journal of Urban and Regional Research* 24, no. 4: 737–757. https://doi.org/10.1111/1468-2427.00276

Urry, John. 2007. *Mobilities.* Cambridge: Polity Press.

Zavestoski, Stephen and Julian Agyeman (eds.). 2014. *Incomplete Streets: Processes, Practices, and Possibilities.* London: Routledge.

Introduction

Tobias Kuttler and Massimo Moraglio

Abstract

In this Introduction, the editors of the volume present its structure and address the theoretical background to the book. After defining the role of mobility in contemporary societies, this Introduction outlines the relationship between transport poverty and social exclusion. The authors demand a better definition and terminology of the subject, claiming that "mobility" poverty addresses the concepts better and in a sounder way. Finally, the Introduction calls for a closer cooperation between the different actors, with a stronger focus on the users of transport.

Recent crises

Recent crises – whether they are at global, regional or local level – demonstrate that *im/mobilities* are at the core of many conflicts. The ongoing global refugee crisis is a constant exercising of the right and ability to move and a demonstration of power to prevent and interrupt such movement. Conflicts over the ability to move can also inflame tensions about larger social, economic and political divides within societies. Questions of mobility seem to steer new and often surprising collaborations among those who consider themselves marginalised and excluded from dominant development discourses, such as the *gilets jaunes* protests in France or the contentious ban on diesel cars in inner cities of Germany. Such struggles do not always employ clearly "progressive" or "reactionary" narratives to make sense of this exclusion, or to envision ways out. Rather, they can be interpreted as a collective or individual reaction to an increasingly hypermobile world where seamless movement seems to be possible for everyone – except for oneself (or one's peers). Such exclusions can be traced back to divisions of class, gender, race, (dis)ability, age and geographical location. However, recent insurgencies are hardly only along these lines. An individual or collective ability to move – or more precisely the potential to individually or collectively decide whether, when and how to move or to stay put – is the result of a combination of different privileges and forms of capital that are put into action in various ways in such decision-making. Conversely, many different forms and processes of discrimination, marginalisation, exclusion and invisibility result in the inability to decide over one's movement or in forced or compelled im/mobility.

Despite pointing out above the contentious nature of mobilities on all scales, this volume has a much more modest scope. It does not analyse the flows and frictions of goods and people on a global scale, nor does it deal with access and control to certain spaces as the "right to the city" entails. This volume is largely about the everyday mobility of people, a field that is usually investigated in transport studies. In everyday mobility, underlying structural disadvantages and privileges are often less obvious and visible in comparison to the immediate deficiencies of transport systems. Such underlying structural disadvantages often appear in very subtle and qualitative ways: locally and temporarily, they take effect in different forms that may obfuscate common roots. Hence, in this volume we will investigate forms of structural disadvantage identified in mobility studies and adapt them more thoroughly to analyse the challenges of everyday mobilities. Therefore, we put forward the concept of *mobility poverty*. This investigation will not be – and can never be – detached from an analysis of systemic privilege, both in a historical and contemporary perspective. The recent COVID-19 crisis shows that the relationship between privilege and bodily movement is currently being recalibrated. Those who are usually highly mobile were among the first to stay at home and work remotely. At the same time, they are delegating the work that requires physical movement to the less fortunate workers of the service industry, either those who are self-employed or on short-term contracts. This means that a degree of physical mobility does not alone exhibit privilege and status (any more). In the light of the COVID-19 experience, perhaps whether network capital – as John Urry coined it – still requires co-presence may also be questioned again. It is the decision about whether, when and how to move and not to move that constitutes privilege today. These are the questions that have been investigated under the mobilities paradigm and under the umbrella of the "politics of mobility" (see e.g. Cresswell 2010).

In the last ten years, debates around equity and justice in transport provision and accessibility have become more dynamic in academic as well as policy circles (see e.g. Currie 2010; Delbosc and Currie 2011; Martens 2012; Martens, Di Ciommo, and Papanikolaou 2014; Martens 2017; Pereira, Schwanen, and Banister 2017). Highlighting the need for a broader conceptualisation under the mobilities paradigm, the concept of mobility justice has gained attention recently (Cook and Butz 2018; Sheller 2018; see also Chapter 1 of this volume). Therefore, there is an urgent need to focus on the spatial and social conditions of mobility inequity and injustice. With a fine-grained analysis of the social and spatial conditions, first, we aim to better conceptualise *mobility poverty* beyond *transport poverty*. Second, we aim to bridge the gap between the concept and its real-world application, by highlighting experiences of mobility poverty through case studies from different European regions. Third, we thereby aim to enhance the expertise of decision-makers and practitioners by offering lessons learnt and recommendations.

Understanding the significance of uneven mobilities

Studying mobilities is not about movements and flows alone. Mimi Sheller summarises the relationship between mobility and immobility very poignantly:

> In contrast to [...] illusions of clean, quick, ethereal mobility, actual mobilities are full of friction, viscosity, stoppages, and power relations. We need to understand not only what is constituted as mobile, or potentially mobile, and what is not, but also where, when, and how there are resistances to that power, or counter-movements against it. Mobilities are always contingent, contested, and performative. Mobilities are never free but are in various ways always channeled, tracked, controlled, governed, under surveillance and unequal—striated by gender, race, ethnicity, class, caste, color, nationality, age, sexuality, disability, etc., *which are all in fact experienced as effects of uneven mobilities.*
> (Sheller 2018, 10, emphasis in original)

Ever since the foundational text of the *new mobilities paradigm* (Hannam, Sheller and Urry 2006), uneven mobilities have been a key topic in mobility studies. Such studies have been heavily influenced by geographers and anthropologists who put forward that the production and creation of spaces, places and communities are processes of establishing links and connections in the form of the movement of bodies, commodities and information. An early influential example is the theory of the social production of space, where Henri Lefebvre points out the significance of networks, flows, connections and linkages in the spatial production process. Space is constituted by networks and it is modified and transformed by networks (Lefebvre 1991, 345–347). When he describes his understanding of spatial practice, he highlights that the relationship between *separation* – e.g. places of residence, work and leisure that are set apart – and *linkages* – e.g. transport infrastructure – is characteristic of the production of space and everyday life, a relationship that is infused with power and hence of a political nature:

> [Spatial practice] embodies a close association [...] between daily reality (daily routine) – and urban reality (the routes and networks which link up the places set aside for work, 'private life' and leisure). This association is paradoxical, because it induces the most extreme separation between the places it links together. 'Modern' spatial practice might thus be defined by the daily life of a tenant in a government-subsidized

high-rise housing project. Which could not to be taken to mean that motorways or the politics of air transport can be left out of the picture.

(Lefebvre 1991, 38)

Also drawing on a relational understanding of space and place, Doreen Massey highlighted that differentiated mobility is a source of power and control:

Mobility, and control over mobility, both reflects and reinforces power. It is not simply a question of unequal distribution, that some people move more than others, and that some have more control than others. It is that the mobility and control of some groups can actively weaken other people. Differential mobility can weaken the leverage of the already weak. The time-space compression of some groups can undermine the power of others.

(Massey 1994, 150)

Especially in the phase of global economic restructuring after 1990, the analysis of networks, flows and linkages – and their disruptions, exclusions and disconnections respectively – became central to understanding the differentiated and unequal outcomes of globalisation. Manuel Castells famously identified two different logics: the *space of flows* – means that the "material arrangements allow simultaneity of social practices without territorial contiguity" – and the *space of place* – understood as a "locale whose form, function, and meaning are self-contained within the boundaries of territorial contiguity" (Castells 1999, 295–296). He argued that:

[...] [T]he constitution of the space of flows [is] in itself a form of domination, since the space of flows, even in its diversity, is interrelated and can escape the control of any locale, while the space of places is fragmented, localized, and thus increasingly powerless vis a vis the versatility of the space of flows. The only chance of resistance for localities is to refuse landing rights for overwhelming flows – only to see that they land in the locale nearby, therefore inducing the bypassing and marginalization of rebellious communities.

(Castells 1999, 297)

Anthropology and geography fertilised each other by highlighting the importance of im/mobilities for a relational understanding of place and community. Considering spatial hierarchies and spatial (dis)connections in social and cultural analysis opened up rich pathways for a better understanding of cultures and societies, an understanding that is sensitive to the questions of power and hegemony. Gupta and Ferguson pointed out: "If one begins with the premise that spaces have always been hierarchically interconnected, instead of naturally disconnected, then cultural and social change becomes not a matter of cultural contact and articulation but one of rethinking difference through connection" (Gupta and Ferguson 1992, 8).

Im/mobilities and disconnections have also been the focus of attention of those who recognised the differentiated effects of infrastructures. Acknowledging the channelling effects of infrastructures, it becomes clear that elements that are physically close are not necessarily connected to each other, while, conversely, elements that are physically distant can indeed be connected. To illustrate this point, Bruno Latour states: "an Alaskan reindeer might be ten meters away from another one and they might be nevertheless cut off by a pipeline of 800 miles that make their mating for ever impossible" (Latour 1996, 372). Also Susan Leigh Star argues that the notion of universally accessible infrastructure is blurred "when one begins to investigate large-scale technical systems in the making, or to examine the situations of those who are not served by a particular infrastructure. [...] One person's infrastructure is another's topic, or difficulty" (Star 1999, 380).

This brief (and incomplete) overview illustrates that investigating the role of im/mobilities and movements and their unequal social consequences has been a topic of many research disciplines for quite a while even before the turn of mobilities. However, before John Urry's *Sociology Beyond Societies* and Zygmunt Bauman's *Liquid Modernity*, mobilities themselves were rarely considered a key epistemological concept in themselves, or as Mimi Sheller puts it, the achievement of the new mobilities paradigm is

> their radical emphasis on *complex mobilities* of all kinds as the ontological basis for all forms of relational space, and partly their deeper cultural analysis of how these political economic relational spaces were produced in and through social and cultural practices.
>
> (Sheller 2018, 11–12, emphasis in original)

Transport poverty and social exclusion

Partially in parallel with, but also in contrast to, advances in social sciences around the new mobilities paradigm, academics in transportation research investigate disadvantages regarding access to transport options and problems in accessibility to locations and opportunities (Figure 0.1). This body of work is often subsumed under the term "transport poverty". Such works have made a substantial contribution by highlighting the reinforcing circle of transport disadvantage, social disadvantage and social exclusion.

Unlike the case of mobilities, the debate around transport poverty was not initially sparked by conceptual and theoretical considerations. Rather, it was fuelled by the necessity for policy-makers and practitioners to deal with the processes of social exclusion, in the light of widening social inequalities in European societies.

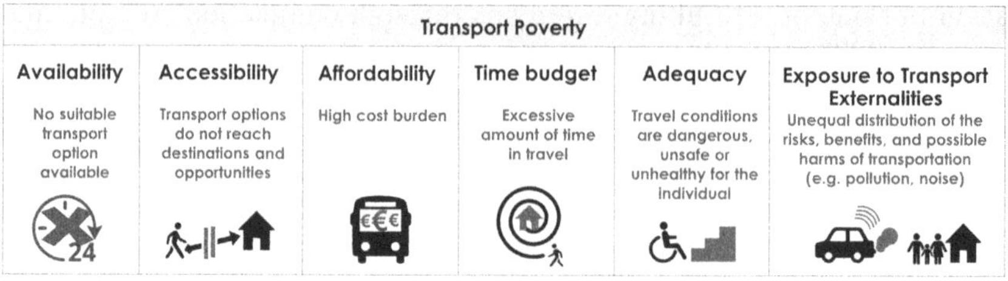

Transport Poverty					
Availability	Accessibility	Affordability	Time budget	Adequacy	Exposure to Transport Externalities
No suitable transport option available	Transport options do not reach destinations and opportunities	High cost burden	Excessive amount of time in travel	Travel conditions are dangerous, unsafe or unhealthy for the individual	Unequal distribution of the risks, benefits, and possible harms of transportation (e.g. pollution, noise)

Figure 0.1 Transport poverty.
Source: Authors, based on Lucas et al. 2016.

Transport disadvantage first received closer attention when the Social Exclusion Unit (SEU) was implemented in the United Kingdom in 1997 as part of the Office of the Deputy Prime Minister, providing strategic advice and policy analysis to the UK government. One explicit focus of the SEU was the role of transport and mobility in processes of social exclusion. In its 2003 report, the SEU pointed out that access to transport options – both individual modes of transport and public transport – may be the result of social exclusion and that inadequate transport availability and accessibility of basic services can reinforce social exclusion (Social Exclusion Unit 2003, 1). It highlighted that social disadvantage and unequal access to transport options can result in a downward spiral that traps individuals, households and social groups in a state of immobility (irrespective of living in urban, peri-urban and rural areas). The report highlighted the inter-linkages between social, spatial and transport-related disadvantage as one of the main factors of social exclusion. Another achievement of the study was that it successfully directed attention to the needs of individuals from vulnerable social groups as well as urban and rural spaces characterised by deprivation and neglect (Church, Frost, and Sullivan 2000; Lucas, Grosvenor, and Simpson 2001; Kenyon, Lyons, and Rafferty 2002). The 2003 report was thus a turning point in the awareness of transport-related disadvantage, generating a growing policy and academic interest in transport-related social exclusion. Today, it is acknowledged that approaches to understanding transport disadvantage and social exclusion must depart from an observation of general poverty towards a contextual understanding of an individual's position within society (Lucas 2012, 106).

The term transport poverty first appeared when it was emphasised in social exclusion literature (Lucas 2004, 1) and became further prominent when it was utilised as a campaigning instrument to shed light on the affordability of car ownership, public transport costs and lack of access to transport (Sustrans 2012). The term has subsequently been used to conceptually clarify the inter-linkages between transport-related disadvantage, social disadvantage, (in)accessibility and social exclusion (Lucas 2012, 107; Titheridge et al. 2014, 19–22).

As described above, transport poverty was initially investigated – to a larger extent – in the context of the debate on social exclusion, equality of access to opportunities and transport justice. In 2012, Karen Lucas made an influential attempt to locate transport poverty at the intersection of transport disadvantage and social disadvantage, the consequences leading to inaccessibility and social exclusion:

> Transport disadvantage and social disadvantage interact directly and indirectly to cause transport poverty. This in turn leads to inaccessibility to essential goods and services, as well as 'lock-out' from planning and decision-making processes, which can result in social exclusion outcomes and further social and transport inequalities will then ensue.
>
> (Lucas 2012, 106)

This statement explains the circular dynamic of production and reinforcement of the experience of disadvantage and social exclusion. This also makes clear that social/transport-related disadvantage and social exclusion are not in a simple causal relationship nor are they synonymous. As Currie and Delbosc point out, when transport disadvantage is explored across high-mobile and low-mobile groups, disadvantage is subjectively estimated very differently (Currie and Delbosc 2010).

The term transport poverty has been continuously employed in more recent years (Velaga et al. 2012, 110; Martens 2013, 24) and is often used interchangeably with other terms, such as 'accessibility poverty' or 'poverty of access' (Farrington and Farrington 2005, 3; Martens and Bastiaanssen 2014, 6–7), 'transport disadvantage' (e.g. Currie et al. 2009, 97–98), 'transport-related' or 'transport-based social exclusion' (e.g. Preston and Rajé 2007, 152–154; Schwanen et al. 2015, 123–125), 'social equity', 'fairness' and 'justice in transport' (e.g. Martens 2009, 4–6; Jones and Lucas 2012, 9; Sheller 2015, 86) and 'transport wealth' (Stokes and Lucas 2011, 4–7). Although these terminologies and the related concepts are defined differently, they also have substantial overlaps and are sometimes based on similar approaches and assumptions. In other words, transport poverty can be approached from different, however, interrelated perspectives (Lucas et al. 2016, 2–4).

In terms of perspectives, transport poverty can be approached from the perspective of *transport affordability*. The term refers to the "financial burden households bear in purchasing transportation services, particularly those required to access basic goods and activities such as healthcare, shopping, school, work and social activities" (Litman 2016, 5). This discourse is largely relevant for industrialised countries, as it centres on ownership of a car as a basic household need. Focusing on affordability to define transport poverty, Gleeson and Randolph (2002, 102) state that "transport poverty occurs when a household is forced to consume more travel costs than it can reasonably afford, especially costs relating to motor car ownership and usage". Several authors have pointed out that forced car ownership[1] is a form of transport poverty that has been identified in the United Kingdom, United States and Australia (e.g. Currie and Senbergs 2007, 2–3).

Another strand of research on transport poverty focuses on accessibility and *accessibility poverty*. Accessibility considers whether people can reach "key services at reasonable cost, in reasonable time and with reasonable ease" (Social Exclusion Unit 2003, 1) – key services primarily being employment, education, health care and daily supply. Accordingly, Martens and Bastiaanssen (2014, 5) define accessibility poverty in the following way: "Accessibility poverty refers to a situation of low accessibility that severely restricts a person's ability to participate in the activities deemed normal in a particular society".

The concept of transport poverty has been criticised from various angles. Firstly, academic debate still lacks an accurate description of the phenomenon of transport poverty and discussion often focuses on how transport poverty differs from poverty itself and who is affected. Due to many open questions, instruments to measure transport poverty are not sufficiently developed, let alone efficient and effective measures to tackle transport poverty. So far, transport poverty has not been taken up by policy-makers as a stand-alone issue that needs to be addressed. Overall, "transport poverty is an extremely under-explored and poorly articulated problem even within developed countries" (Lucas et al. 2016, 10).

Secondly, transport poverty is often discussed in the context of the endeavour to build inclusive societies. However, how exactly and to what degree transport and mobility-related disadvantages contribute to social exclusion, reduced opportunities and well-being are still insufficiently explored. Such insufficient knowledge can be partially traced back to the weaknesses of the social inclusion/exclusion concept. The concept often lacks clarity with the result that "social inclusion is conceived in many alternative ways, depending upon ideology" (Silver 2015, 3–4). Due to the wide use of the term often without a substantial analytical framework, Kasper et al. (2017, 7) conclude that social inclusion has become one of the many aspirational terms that "tend to lose any precision to their meaning, and come to be good by definition rather than by implication." It is also pointed out that the aim of social inclusion itself can be contested. Schwanen et al. (2015) in an important contribution remind us that social capital (in Pierre Bourdieu's understanding of social capital) has both inclusionary and exclusionary effects. Hence, they argue that "fully appreciating the Janus-faced character [...] helps us understand the dynamics in the interactions between mobility and social exclusion because it both is a medium for social change and can reinforce existing inequalities." Hilary Silver goes a step further when she highlights that the idea of social inclusion is infused with power: "Paradoxically, recognizing and assisting an excluded group in the name of inclusion may simultaneously stigmatize, label, or include them in ways they did not choose [...]. Inclusion sounds good – but on whose terms?" (Silver 2015, 22).

These criticisms should be taken seriously when transport disadvantages are investigated in order to create inclusive societies. In this sense, it is important to keep in mind that social exclusion is a multi-layered problem that has many causes other than transport-related disadvantages (Currie and Delbosc 2010, 964). Furthermore, as others have also argued, transport disadvantage and transport-related social exclusion are not synonymous with each other: it is possible to be socially excluded but still have good access to transport or to be transport disadvantaged but highly socially included (Currie and Delbosc 2010; Lucas 2012).

Towards mobility poverty

The new mobilities paradigm emphasises mobilities over transportation and clarifies the differences between the two. However, the terms "mobility poverty" and "transport poverty" are sometimes used interchangeably. Unlike the international literature on the topic, the German academic debate more often explicitly differentiates between mobility poverty (*Mobilitätsarmut*) und transport poverty (*Vekehrsarmut*) (Dangschat 2011; Daubitz 2016; Stark 2017; Schwedes et al. 2018). Schwedes et al. (2018, 79) stress the point that it is not only the spatial delimitations that influence mobility behaviour, but also the "mental horizon" and the capacity to plan and shape one's own life. Mobility is therefore also mental flexibility and agility and the created personal sphere is one's "space of opportunity". These arguments, very much related to Vincent Kaufmann's concept of motility (Kaufmann, Bergman, and Joye 2004), call for an analysis of "mobility poverty" and not "transport poverty". According to Schwedes et al. (2018), transport poverty focuses too much on the availability of transport options and that, even in the most sophisticated and well-grounded works, the topic is mainly accessed from the perspective of transport provision.

Furthermore, it can be observed that in transport poverty literature, different phenomena of social disadvantage have been explored with unequal attention; while there is substantial empirical evidence on the link between material poverty and transport disadvantage, there is less knowledge about the role of other socio-demographic features that are experienced as disadvantage. This includes gender, disability, old age and young age. There is far less attention on the discrimination that ethnic minorities, migrants and refugees face while being mobile and accessing transport (see e.g. Rajé 2017).

Finally, despite significant progress, the relationship between spatial and temporal organisation in daily life and experienced disadvantage is still insufficiently explored under the umbrella of transport poverty. The challenge of assessing this topic is related to the *relational nature of high mobility and low mobility*: increasing overall mobility levels – and instances of high mobility in certain sections of society – may have the consequence of reduced mobility and accessibility of those already less mobile in societies. Consequently, "what is necessary for full 'social' inclusion varies as the means and modes of mobility change and as the potential for 'access' develops" (Cass, Shove, and Urry 2005, 542). The highly contextual and relational nature of mobility disadvantage needs further exploration.

Hence, we turn towards examining *mobility poverty*. Mobility poverty focuses on barriers to people's ability and potential to move. Unlike the accessibility perspective that aims foremost at examining basic needs, mobility poverty takes into perspective the effects of hypermobility and the development of highly mobile societies. In such a context, due to the power relations involved, especially vulnerable groups experience reduced mobility options and accessibility levels (Massey 1994, 150; Kenyon, Lyons and Rafferty 2002, 210; Cass, Shove and Urry 2005, 542; Hannam, Sheller and Urry 2006, 3). Mobility perspectives take factors at micro-, meso- and macro-level into account, such as status, wealth, prestige and power, and highlight that mobility is fundamentally linked to social, cultural, economic and political processes. Key to the approach is the identification of the *systemic* lack of transportation and mobility options and the *relational* nature of transport problems. This mobility framing is not only sensitive to questions of access, but also to the skills and capabilities of individuals as well as to personal ambitions and differentiated needs. By understanding mobility needs as socially constructed, it explores the gap between unrealised mobility needs and actual travel patterns. The key concept to such an in-depth appraisal of differentiated mobilities is the *motility* approach by Vincent Kaufmann that will be presented in Chapter 1 (Kaufmann, Bergman, and Joye 2004; Flamm and Kaufmann 2006).

Similarly, John Urry directed our attention to the dynamic, ever-changing and complex prerequisites for a meaningful social life. Turning away from the "formal" aspects that are addressed in the accessibility perspective – employment, education, health – to the social relationships of everyday life, he points out that it is actually the link between social relationships and mobility that creates new inequalities. These inequalities imply additional rewards (also in material terms) for some, and new burdens for others. He therefore developed the concept of network capital,

> the capacity to engender and sustain social relations with those people [who] are not necessarily proximate and which generates emotional, financial and practical benefit [...]. Network capital [...] produce[s] a distinct stratification order that now sits alongside social class, social status and party.
>
> (Urry 2007, 197)

Such elaborated analyses of mobility disadvantage pose a considerable challenge for investigating the phenomena both on a representative and comparable scale, posing a problem for national and local policy-makers as to how to adequately address the issue.

As one of the first, Lucas et al. (2016) have explicitly outlined *mobility poverty* as one form of how transport poverty can be experienced. Understanding mobility poverty as a sub-phenomenon of transport poverty is reasonable, since the term transport poverty effectively separates the topic of everyday and regular corporeal movement in space from other forms of movement such as migration, which are not the direct focus of transportation researchers.

Similar to Karen Lucas' area of research, this publication also focuses mostly on everyday travel. *However, we propose that transport poverty should rather be understood as one of many forms of how mobility poverty can be experienced.* This shift in analysing mobility-related disadvantages puts mobilities in the context of many other processes and dynamics. Although the international transport debate has made considerable progress in integrating aspects of mobility poverty into the debate of transport poverty, transport academics often hesitate to make explicit the systemic unevenness of mobilities and spaces. This unevenness is however central to a holistic understanding of all kinds of mobilities, as Sheller (2018) highlights. We believe that the consideration of unevenness of mobilities should also inform a more profound analysis of everyday and regular corporeal movement, which is the objective of this volume. Following this understanding, in this volume we consciously make the shift from *transport poverty* to *mobility poverty*.

Mobility poverty is a phenomenon that can appear anywhere and at any point in time. Indeed, mobility poverty – as will be shown – is highly relational and contextual. It is by no means synonymous with material poverty. Although mobility poverty is often linked to material poverty – and other forms of mobility disadvantages may be reinforced by material poverty – mobility poverty can appear in personal conditions of wealth or in economically dynamic regions. Forms of mobility can be most disruptive and, at the same time, the least visible under exactly such circumstances. The following chapters, and especially the experiences from the field (Part IV), will further illuminate this proposition.

Even after making this shift to mobility poverty, the joint objective must be the development of effective and applicable solutions, despite differentiated mobility needs in a complex world. In an ever more complex and growing field, there are many barriers that prevent policy-makers and practitioners from taking positive action because it is difficult to identify the right level and scale of intervention. The actual points of intervention that allow the multidimensional, interrelated and always context-specific challenges to be tackled in efficient and effective ways have not yet been precisely identified. This volume aims to fill this gap.

Selection of the fieldwork study regions and targeted vulnerable groups

The selection of study regions was result of an in-depth analysis of geographical, socio-economic and other negative factors (i.e. deprived, dispersed or rural peripheral), transport and accessibility problems and running experiences and initiatives. Thereby, it was not only the aim to identify the most deprived or marginalised regions in Europe, or those with lowest accessibility, but rather to **identify a set of case study regions that present a variety of different spatial, social and mobility characteristics that occur in Europe** (Figure 0.2; Table 0.1).

Furthermore, the selection process considered that mobility disadvantage is a multifaceted phenomenon appearing in both beneficial and disadvantaged socio-economic environments. It was taken into account that it is possible that on the very micro-level individuals subjectively experience a form of mobility disadvantage and subsequent social exclusion even with high socio-economic status. Also, it was assumed that perception of disadvantage and marginalisation can even be stronger in dynamic and prosperous regions characterised by high mobility levels. Hence, the selection process followed the understanding that socio-economic disadvantage (most importantly absolute/relative poverty) is only one of the indicators for mobility poverty. Consequently, mobility poverty on the micro-level was also analysed in comparatively better-off regions.

Through such a variety, it is possible to relate – cautiously – to similar cases and situations in other parts of Europe, and by theoretical inference draw general conclusions while being sensitive to the local context. Thus, while the case studies do not claim representativeness for the respective countries, the particular European region or the European Union as a whole, the study methodology nevertheless allows developing a better picture of mobility poverty in Europe.

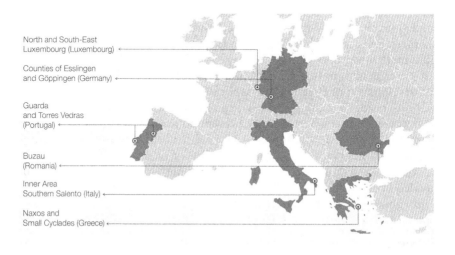

Figure 0.2 Fieldwork study regions.
Source: HiReach Project.

Table 0.1 Fieldwork study regions with spatial layers and vulnerable groups addressed

COUNTRY	AREA	GEO LAYER	SOCIAL LAYER						
			CHILDREN	ELDERLY PEOPLE	PEOPLE WITH REDUCED MOBILITY	WOMEN	MIGRANTS AND ETHNIC MINORITIES	LOW INCOME AND UNEMPLOYED	POPULATION LIVING IN REMOTE RURAL AND DEPRIVED AREAS
Germany	Counties of Esslingen and Goppingen	Peri-urban with mixed rural and urban		●		◉	●		
Greece	Naxos and Small Cyclades	Remote rural	●		◉				●
Italy	Inner Area Southern Salento	Remote peri-urban		◉	●	●	●		◉
Luxemb.	North and south-east Luxembourg	Peri-urban and rural					◉	●	●
Portugal	Guarda	Urban and remote rural		●				●	◉
Romania	Buzau	Urban/Sub-urban	●	●	●			●	●

Main target ●

Secondary target ◉

Source: HiReach Project.

Study regions have been assessed according to the following criteria:

1 **Targeted vulnerable social groups:** presence and opportunities to involve and mix mobility needs of different user groups, namely:

- Low income and unemployed people
- Elderly people
- People with reduced mobility
- Women
- Migrants and ethnic minorities
- Children and young people
- People living in remote, rural or deprived urban areas.

2 **Assessment of socio-demographic negative factors:** e.g. poverty indicators (absolute/relative), GDP per capita, purchasing power, unemployment rate, ageing ration, education levels, access to health services, access to goods and services, housing, gender imbalances.

3 **Targeted geographical characteristics:** rural, peri-urban, urban.

4 **Assessment of geographical negative factors:** remote areas, dispersed and/or scattered settlement structure, topography.

5 **Characteristics of the transport and mobility system:** modal split, motorisation, road network, public transport coverage, service levels of public transport, accessibility and barrier-free design of public transport.

6   **Ongoing initiatives:** volunteer services, sharing/peer-to-peer models, integrated mobility platforms, public/private DRT services.
7   **Key stakeholders:** groups, associations, NGOs, local experts or public/private transport operators.
8   **Other aspects:** information on beneficial or unfavourable aspects, for example, start-ups being active in the field, social innovation initiatives.

Fieldwork methodology

During desk research on mobility poverty, knowledge gaps and additional topics were identified that were taken up and explored more deeply during the fieldwork phase. Furthermore, findings from desk research needed to be verified and contextualised in the field. Hence, six study regions were identified at the beginning of the HiReach project, as outlined above. In these study regions, research with experts and members of different social groups was conducted, in order to arrive at an in-depth understanding of mobility and accessibility problems experienced in different countries at urban, peri-urban and rural level. The fieldwork methodology relied on interviews with experts and on focus group discussions with members of different social groups and stakeholders.

In a first step, in each study region, **interviews were conducted with representatives of local authorities, transport operators, non-governmental organisations and interest groups** (e.g. representing persons with disabilities, unemployed persons, children, elderly, migrants, women and other vulnerable to exclusion citizens). These interviews had the objective to identify social as well as mobility-related disadvantages in each local study region.

The second step was the conduction of **focus group sessions with members of social groups**, to further validate and explore their mobility and accessibility challenges. The focus group sessions took place in two rounds, in summer to autumn 2018 and summer 2019. The first round in 2018 focussed on identifying challenges and problems with personal mobility. In 2018, two focus groups sessions per targeted social group in each of the study regions were conducted, which means four focus group sessions per study region. The second round in 2019 aimed at identifying and developing solutions. Two focus group sessions in each study region were conducted in 2019. In many cases, the second round of focus group was conducted in the same municipality, in collaboration with the same local organisation. Also, in some cases, participants from the 2018 sessions joined again in 2019, which was ideal to discuss a possible pathway from identifying challenges to preparing solutions.

In each focus group session, between 6 and 12 persons participated. A gender balance was aimed at in these sessions, however not always achieved. In some cases, it was deemed beneficial to conduct separated sessions for men and women, in order to be able to identify needs of women more thoroughly. This was, for example, the case for the focus group sessions with refugees in Esslingen/Germany. The focus group sessions allowed us to discuss the previously identified problems from different, sometimes surprising perspectives with the participants. Apart from verifying existing knowledge, new issues were raised or linkages to other aspects of everyday life were drawn that the researchers did not imagine earlier.

The focus groups were conducted in close collaboration with local NGOs, authorities and advocacy groups. Participants were recruited via the channels of these organisations. Participation in the focus group session was incentivised by shopping vouchers, and all expenses for transport were refunded to the participants. Child care was provided for children of participants where necessary.

Research activities with members of different social groups were conducted according to the European and national legislation in force concerning research ethics and vulnerable groups of citizens and the requirements for informed consent. Special attention had to be given to preventing the risk of enhancing vulnerability and stigmatisation of individuals, especially for the most vulnerable groups. In this respect, involvement of children, migrants and refugees had to be treated with utmost caution. An ethics board supervised data collection, focus group methodology and involvement of the target groups.

Structure of the volume

This book is divided into four parts, organised around four thematic areas:

Part I focuses on social skills and individual aptitudes on the one hand and mobility justice on the other. This part devotes three chapters to the social arrangement of mobility, switching between the social and individual levels. Chapter 1 is devoted to "Learning Mobility" while Chapter 2 deals with "Unequal mobilities, network capital and mobility justice". Chapter 3 ends this section by dealing with "The impact of life events on travel behaviour".

Part II deals with spatial elements of mobility poverty. Chapter 4 ("The spatial dimension of mobility poverty") offers initial insights into spatial systems in Europe and their effects on mobility disadvantage, while Chapter 5 addresses "The urban arena" and Chapter 6 "the rural arena". Together, the three chapters aim to define the characteristics of the urban, peri-urban and rural areas and how spatial contexts shape mobility poverty.

Part III tackles the societal roots and impacts of mobility poverty. The social component of mobility poverty is addressed analysing, for each of the six chapters forming this part, different vulnerable groups, for example: Chapter 7, "Women and gender-related aspects"; Chapter 8, "People on low income and unemployed persons"; Chapter 9, "Impacts on mobility in an ageing Europe"; Chapter 10, "The predicaments of European disabled people"; Chapter 11, "Migrants, ethnic minorities and mobility poverty"; and Chapter 12, "Children and young people".

Part IV covers the results of the theoretical enquiry to the field. In this final part, the investigation of six case studies as carried out in Portugal (Chapter 13), Italy (Chapter 14), Greece (Chapter 15), Romania (Chapter 16), Germany (Chapter 17) and Luxembourg (Chapter 18) is offered to the reader, including the outcomes of the focus groups and users' engagement. The conclusion ends this volume.

This book has been made possible by the continuous engagement of many people. First of all, we should mention the many authors of the chapters, but also those working on the Horizon 2020 project HiReach, on which this book is based. We are also very grateful to the HiReach advisory committee of practitioners ("Take-up-Group"), which accompanied the project, and, naturally, to those stakeholders who supported the project who have been instrumental in enabling us to conduct the fieldwork. The engagement of local actors, the voices of the users, made it possible to test and calibrate our project and its results. We think that this book should be dedicated to those who encounter problems in their everyday mobility.

Note

1 The low income of households, the need for travel in everyday life and the lack of alternatives to the car in the form of public transport are central to the concept of forced car ownership. In such a context, expenditure on a car is seen as essential and unavoidable in the household concerned. Hence, there is an incidence of high car ownership, but at high cost to low income groups (see Currie and Senbergs 2007, 2).

References

Cass, Noel, Elizabeth Shove, and John Urry. 2005. "Social exclusion, mobility and access." In *The Sociological Review* 53 (3): 539–555. https://doi.org/10.1111/j.1467-954X.2005.00565.x.

Castells, Manuel. 1999. "Grassrooting the space of flows." In *Urban Geography* 20 (4): 294–302. https://doi.org/10.2747/0272-3638.20.4.294.

Church, Andrew, Martin Frost, and Karen Sullivan. 2000. "Transport and social exclusion in London." In *Transport Policy* 7 (3): 195–205. https://doi.org/10.1016/S0967-070X(00)00024-X.

Cook, Nancy, and David Butz. 2018. *Mobilities, Mobility Justice and Social Justice.* London: Routledge.

Cresswell, Tim. 2010. "Towards a politics of mobility." In *Environment and Planning D: Society and Space* 28 (1): 17–31. https://doi.org/10.1068/d11407.

Currie, Graham. 2010. "Quantifying spatial gaps in public transport supply based on social needs." In *Journal of Transport Geography* 18 (1): 31–41. https://doi.org/10.1016/J.JTRANGEO.2008.12.002.

Currie, Graham, and Alexa Delbosc. 2010. "Modelling the social and psychological impacts of transport disadvantage." In *Transportation* 37 (6): 953–966. https://doi.org/10.1007/s11116-010-9280-2.

Currie, Graham, Tony Richardson, Paul Smyth, Dianne Vella-Brodrick, Julian Hine, Karen Lucas, Janet Stanley, Jenny Morris, Ray Kinnear, and John Stanley. 2009. "Investigating links between transport disadvantage, social exclusion and well-being in Melbourne—Preliminary results." In *Transport Policy* 16 (3): 97–105. https://doi.org/10.1016/j.tranpol.2009.02.002.

Currie, Graham, and Zed Senbergs. 2007. "Exploring forced car ownership in metropolitan Melbourne." *Proceedings of the 30th Australasian Transport Research Forum.* https://research.monash.edu/en/publications/exploring-forced-car-ownership-in-metropolitan-melbourne. Accessed 21 April 2020.

Dangschat, Jens S. 2011. "Social cohesion-eine Herausforderung für das Wohnungswesen und die Stadtentwicklung?" In *Forum Wohnen und Stadtentwicklung* (1/2011): 1–8. https://www.vhw.de/fileadmin/user_upload/08_publikationen/verbandszeitschrift/2000_2014/PDF_Dokumente/2011/FWS_1_2011/FWS_1_11_Dangschat.pdf. Accessed 17 March 2020.

Daubitz, Stephan. 2016. "Mobilitätsarmut: Die Bedeutung der sozialen Frage im Forschungs-und Politikfeld Verkehr." In *Handbuch Verkehrspolitik*, edited by Oliver Schöller, Weert Canzler and Andreas Knie, 433–447. Wiesbaden: Springer.

Delbosc, Alexa, and Graham Currie. 2011. "Using Lorenz curves to assess public transport equity." In *Journal of Transport Geography* 19 (6): 1252–1259. https://doi.org/10.1016/J.JTRANGEO.2011.02.008.

Farrington, John, and Conor Farrington. 2005. "Rural accessibility, social inclusion and social justice: Towards conceptualisation." In *Journal of Transport Geography* 13 (1): 1–12. https://doi.org/10.1016/j.jtrangeo.2004.10.002.

Flamm, Michael, and Vincent Kaufmann. 2006. "Operationalising the concept of motility: A qualitative study." In *Mobilities* 1 (2): 167–189. https://doi.org/10.1080/17450100600726563.

Gleeson, Brendan, and Bill Randolph. 2002. "Social disadvantage and planning in the Sydney context." In *Urban Policy and Research* 20 (1): 101–107. https://doi.org/10.1080/08111140220131636.

Gupta, Akhil, and James Ferguson. 1992. "Beyond 'culture': Space, identity, and the politics of difference." In *Cultural Anthropology* 7 (1): 6–23. https://doi.org/10.1525/can.1992.7.1.02a00020.

Hannam, Kevin, Mimi Sheller, and John Urry. 2006. "Mobilities, immobilities and moorings." In *Mobilities* 1 (1): 1–22. https://doi.org/10.1080/17450100500489189.

Jones, Peter, and Karen Lucas. 2012. "The social consequences of transport decision-making: Clarifying concepts, synthesising knowledge and assessing implications." In *Journal of Transport Geography* 21: 4–16. https://doi.org/10.1016/j.jtrangeo.2012.01.012.

Kasper, Eric, Gordon McGranahan, D. J.H. te Lintelo, Jaideep Gupte, Jean-Pierre Tranchant, R. W.D. Lakshman, and Zahrah Nesbitt-Ahmed. 2017. *Inclusive Urbanisation and Cities in the Twenty-First Century.* Evidence Report No 220. Brighton: Institute of Development Studies. https://opendocs.ids.ac.uk/opendocs/bitstream/handle/20.500.12413/12849/ER220_InclusiveUrbanisationandCitiesintheTwentyFirstCentury.pdf;jsessionid=6F0EA3601961EFA16482C12C556DE920?sequence=1. Accessed 17 March 2020.

Kaufmann, Vincent, Manfred M. Bergman, and Dominique Joye. 2004. "Motility: Mobility as capital." In *International Journal of Urban and Regional Research* 28 (4): 745–756. https://doi.org/10.1111/j.0309-1317.2004.00549.x.

Kenyon, Susan, Glenn Lyons, and Jackie Rafferty. 2002. "Transport and social exclusion: Investigating the possibility of promoting inclusion through virtual mobility." In *Journal of Transport Geography* 10 (3): 207–219. https://doi.org/10.1016/S0966-6923(02)00012-1.

Latour, Bruno. 1996. "On actor-network theory: A few clarifications." In *Soziale Welt* 47 (4): 369–381.

Lefebvre, Henri. 1991. *The Production of Space*. Oxford: Blackwell.

Litman, Todd. 2016. "Transportation affordability." In *Transportation* 250: 360–1560. Updated version: https://www.vtpi.org/affordability.pdf. Accessed 9 April 2020.

Lucas, Karen. 2004. *Running on Empty: Transport, Social Exclusion and Environmental Justice*. Bristol: Policy Press.

Lucas, Karen. 2012. "Transport and social exclusion: Where are we now?" In *Transport Policy* 20: 105–113. https://doi.org/10.1016/j.tranpol.2012.01.013.

Lucas, Karen, Tim Grosvenor, and Roona Simpson. 2001. *Transport, the Environment and Social Exclusion*. York: Joseph Rowntree Foundation.

Lucas, Karen, Giulio Mattioli, Ersilia Verlinghieri, and Alvaro Guzman. 2016. "Transport poverty and its adverse social consequences." In *Proceedings of the Institution of Civil Engineers* 169 (TR6): 353–365. https://doi.org/10.1680/jtran.15.00073.

Martens, Karel. 2012. "Justice in transport as justice in accessibility: Applying Walzer's 'Spheres of Justice' to the transport sector." In *Transportation* 39, 1035–1053. https://doi.org/10.1007/s11116-012-9388-7.

Martens, Karel. 2013. "Role of the bicycle in the limitation of transport poverty in the Netherlands." In *Transportation Research Record: Journal of the Transportation Research Board* 2387 (1): 20–25. https://doi.org/10.3141/2387-03.

Martens, Karel. 2017. *Transport Justice*. New York and Abingdon: Routledge.

Martens, Karel, and Jeroen Bastiaanssen. 2014. "An index to measure accessibility poverty risk." *Paper Presented at Colloquium Vervoersplanologisch Speurwerk*, 21 November 2014, Eindhoven, the Netherlands. https://doi.org/10.1016/B978-0-12-814818-1.00003-2.

Martens, Karel, Floridea Di Ciommo, and Anestis Papanikolaou. 2014. "Incorporating equity into transport planning: Utility, priority and sufficiency approaches." In *Proceedings of the XVIII Congreso Panamericano de Ingeniería de Tránsito, Transporte y Logística*, Santander, Spain. https://www.parking-mobility.org/2020/03/04/mrc-incorporating-equity-into-transport-planning-utility-priority-and-sufficiency-approaches/. Accessed 21 April 2020.

Massey, Doreen. 1994. *Space, Place and Gender*. Cambridge: Polity Press.

Pereira, Rafael H. M., Tim Schwanen, and David Banister. 2017. "Distributive justice and equity in transportation." In *Transport Reviews* 37 (2): 170–191. https://doi.org/10.1080/01441647.2016.1257660.

Preston, John, and Fiona Rajé. 2007. "Accessibility, mobility and transport-related social exclusion." In *Journal of Transport Geography* 15 (3): 151–160. https://doi.org/10.1016/j.jtrangeo.2006.05.002.

Rajé, Fiona. 2017. *Negotiating the Transport System: User Contexts, Experiences and Needs*. Abingdon: Routledge.

Schwanen, Tim, Karen Lucas, Nihan Akyelken, Diego C. Solsona, Juan-Antonio Carrasco, and Tijs Neutens. 2015. "Rethinking the links between social exclusion and transport disadvantage through the lens of social capital." In

Transportation Research Part A: Policy and Practice 74: 123–135. https://doi. org/10.1016/j.tra.2015.02.012.

Schwedes, Oliver, Stephan Daubitz, Alexander Rammert, Benjamin Sternkopf, and Maximilian Hoor. 2018. *Kleiner Begriffskanon der Mobilitätsforschung.* 2nd edition. IVP-Discussion Paper. Berlin. https://www.econstor.eu/bitstream/10419/ 200083/1/ivp-dp-2018-1.pdf. Accessed 21 April 2020.

Sheller, Mimi. 2015. "Racialized mobility transitions in Philadelphia: Connecting urban sustainability and transport justice." In *City & Society* 27 (1): 70–91. https:// doi.org/10.1111/ciso.12049.

Sheller, Mimi. 2018. *Mobility Justice: The Politics of Movement in an Age of Extremes.* London and Brooklyn: Verso Books.

Silver, Hilary. 2015. "The contexts of social inclusion." DESA Working Paper No. 144. https://www.un.org/esa/desa/papers/2015/wp144_2015.pdf. Accessed 17 March 2020.

Social Exclusion Unit. 2003. "Making the connections: Final report on transport and social exclusion." http://www.ilo.org/wcmsp5/groups/public/@ed_emp/@emp_ policy/@invest/documents/publication/wcms_asist_8210.pdf. Accessed 13 June 2018.

Star, Susan L. 1999. "The ethnography of infrastructure." In *American Behavioral Scientist* 43 (3): 377–391. https://doi.org/10.1177/00027649921955326.

Stark, Kerstin. 2017. "Mobilitätsarmut in der sozialwissenschaftlichen Debatte." In *Energie und soziale Ungleichheit,* edited by Großmann, Katrin, André Schaffrin, and Christian Smigiel, 79–100. Wiesbaden: Springer.

Stokes, Gordon, and Karen Lucas. 2011. "National travel survey analysis." In *Transport Studies Unit, School of Geography and the Environment.* https://www. tsu.ox.ac.uk/pubs/1053-stokes-lucas.pdf. Accessed 9 April 2020.

Sustrans. 2012. "Measuring and mapping transport poverty." http://www.sustrans. org.uk/measuring-mapping-transport-poverty. Accessed 17 August 2018.

Titheridge, Helena, Roger L. Mackett, Nicola Christie, Daniel Oviedo Hernández, and Runing Ye. 2014. *Transport and Poverty: A Review of the Evidence.* London: UCL Transport Institute, University College London. https://discovery.ucl.ac.uk/ id/eprint/1470392/1/transport-poverty[1].pdf. Accessed 17 March 2020.

Urry, John. 2007. *Mobilities.* Cambridge: Polity Press.

Velaga, Nagendra R., Mark Beecroft, John D. Nelson, David Corsar, and Peter Edwards. 2012. "Transport poverty meets the digital divide: Accessibility and connectivity in rural communities." In *Journal of Transport Geography* 21: 102–112. https://doi.org/10.1016/j.jtrangeo.2011.12.005.

Part I

Social skills and individual aptitudes

1 Learning mobility

Tobias Kuttler and Massimo Moraglio

Abstract

In this chapter, we take a social constructivist approach to mobility poverty. We argue that, for an in-depth understanding of the phenomenon, it is crucial to investigate the contentious relationship of *mobilities* and *immobilities*. Whether, when and how people decide to move, or to stay immobile, is a complex process. In this chapter, we highlight the aspect of *learning* how to be mobile and at the same time point to the factors that limit the learning process. With such an approach, we lay the groundwork for a better understanding of unmet mobility needs along with the interrelation between realised and unrealised mobilities.

Introduction: realised and unrealised mobility – a social constructivist approach

This and the following chapter seek to depart from a classical account of transport disadvantage focusing on material poverty in order to arrive at a more complex understanding of inequality, disadvantage and injustice, which we subsume under the concept of *mobility poverty*. This considers the increasing variability of lifestyles, attitudes, opinions and values, how they play out on a micro-societal level as well as within the same social stratum, ultimately how this constellation of factors affects people's mobility. We argue that mobility poverty is not only about a lack or shortage of actual movement. It is about the conditions that presuppose actual realised movement such as individual factors that create a desire, motivation and need to be mobile. Furthermore, mobility poverty is not only about differentiated mobility but also about the liberty to move or not move, or the decision to be mobile or stay immobile.

In this vein, the arguments presented here will allow us to (i) better understand the *conjunction between mobility and immobility* and (ii) to identify the *gap between realised and unrealised mobility,* which leads to an investigation of *unmet mobility needs.*

What exactly is our aim by highlighting the gap between realised and unrealised mobility? This is not immediately self-evident, but, for some social groups, this gap is more apparent and can be identified more easily. For example, this gap can be immediately grasped when studying the mobility of the elderly: both anecdotic evidence and research indicate that there are lower levels of mobility activity than people actually desire. Mostly, mobility is desired to foster social relationships and conduct leisure activities, but also to maintain social reputation and access social resources (Hjorthol 2013, 1194). This desire increases with age. It could also be shown that, with increasing age, meeting basic needs like shopping receive wider significance for a person's well-being. For example, this boils down to personal assurance to be independent and in control of one's life, the possibility to meet friends, or just the positive feeling of being out of the house, "on the road" or among people (Hjorthol 2013, 1203–1206). Conversely, the desire for mobility among the elderly often remains unrealised due to inadequate transport options, limited financial means and physical constraints. In addition, often elderly people are reluctant to rely on support from friends and relatives to meet their mobility demands due to internalised norms of self-reliance and independence, hence mobility remains unrealised (Schwanen, Banister, and Bowling 2012, 1320; Ziegler and Schwanen 2011, 777).

In many other cases, the exploration of this mobility gap poses a challenge, which is also eventually a methodological challenge. Research on mobility requirements often relies on the observation of travel that actually takes place. Traditionally with quantitative methods, and increasingly with qualitative or mixed-method approaches, travel behaviour and travel patterns are explored through surveys, travel diaries, GPS tracking, focus groups and in-depth interviews. Although such studies deliver important results on mobility behaviour and patterns and provide a strong basis for transport modelling and demand forecasting, the deeper-lying norms and attitudes of individuals towards their spatial movement often remain hidden. While many studies differentiate between various trip purposes, they do not explore the more fine-grained motivations for being mobile or immobile in one or another way, and thus miss blocked desires (Nordbakke and Schwanen 2015, 1130–1131; Pereira, Schwanen, and Banister 2017, 177).

The definition and identification of mobility desires and unrealised mobility needs could potentially have a strong impact on policy formulation. The way in which mobility needs are defined depends on who participates in the political or agenda setting process. A participative and inclusive process is crucial for the policies and solutions developed. This means that those who do not have access to political decision-making – or are not adequately represented – may not have the chance to express their mobility needs. Especially in the case of socially disadvantaged groups, new policies or solutions often do not address those specific needs or, worse, policies can even further hamper the ability to participate in social life (see e.g. Lucas 2006, 806; Rajé 2007, 66).

In order to analyse the gap between realised and unrealised mobility, and identify unmet mobility needs, we take a social constructivist approach.

Indeed, individual motivations and needs to be im/mobile are socially constructed: the "desire" or "necessity" to move is highly discretional according to social and cultural context. What seems indispensable to one person may be mundane to another. The motivation to move, as well as the ability to move, is closely linked to social norms, values, experiences and socially embedded expectations.

When analysing the gap between realised and unrealised mobility, it becomes necessary to differentiate *individual mobility needs and aptitudes* on the one hand and the notion of *unequal mobilities* and *mobility justice* on the other hand, keeping in mind that for us this goes beyond debates about the distribution of accessibility levels. While these are by no means mutually exclusive approaches, different strands of investigation put varying emphasis on each of them, resulting in different conclusions about what is necessary to achieve what is called a "good" and meaningful life, and subsequent recommendation for policy and technological solutions.

In this chapter, we begin with scrutinising how individual mobility motivations, needs and desires are developed in a setting of established social and cultural norms, values, experiences and socially embedded expectations. While navigating social and cultural settings, *mobility is learnt by individuals in a complex and long-lasting process.* In this learning process, individual mobility aptitudes and skills are established, which enable individuals in their decisions of whether to move or not in quite different ways.

In order to analyse these concepts, we scrutinise:

The role of social networks and the significance of being mobile for social purposes.
The concept of motility in order to shed light on enabling and disabling factors of mobility.
Then, turning more concretely towards the aspect of "learning", we focus on the role of socialisation and the process of "learning mobility".

Fourth, the process of obtaining travel know-how and spatial knowledge will be elaborated.

Lastly, the gap between realised and unrealised mobility will be illustrated by the example of virtual mobilities.

Mobilities and social networks

Especially for groups that are considered vulnerable, the primary attention of decision makers and practitioners is often on securing the basic and formal needs of everyday life: employment, education, health care. However, such a model

> rests on a definition of what excluded people should want or need and obscures the role that social networks play in maintaining a 'good life' and in structuring the meaning of inclusion and participation. [...] This is difficult to achieve, but one method is to focus upon 'blocked desire', especially when people cannot meet what they take to be important obligations of co-presence.
>
> (Cass, Shove, and Urry 2005, 551)

Hence, the importance of maintaining social networks for vulnerable social groups and the associated necessities to be mobile need further scrutiny.

The role of social networks for maintaining a "good life" and the associated need for mobility has been widely discussed. John Urry and other scholars (Urry 2007; Urry and Grieco 2011) have described in detail the significance and even primacy of social relations for maintaining a meaningful life in a networked society:

> What seems important in contemporary life are overlapping and intersecting social networks – in leisure, friendship, family life as well as in work and organizations. And these networks appear to demand intermittent travel, such travel being crucial to forming and sustaining such networks produced through 'moments of co-presence'.
>
> (Cass, Shove, and Urry 2005, 545)

This leads us to conceive that social relations for some groups – especially elderly and mobility-impaired people in rural areas with inadequate public transport – are a prerequisite for being mobile, while mobility of these individuals again reinforces the ability for co-presence and hence the stability of social ties (Jansuwan, Christensen, and Chen 2013 for low-income groups; Lovejoy and Handy 2011 for migrants in the United States; Pyer and Tucker 2014 for young people with disability; Rajé 2007 for poor elderly people; Rittner and Kirk 1995). It has also been shown that different socially constructed needs can be in conflict with each other: many older people are reluctant to rely on support from friends and relatives to meet their mobility demands due to internalised norms of self-reliance and independence, with the effect that especially the elderly tend not to participate in social and cultural life if they would need assistance with transportation from friends and relatives (or technical devices) (Schwanen, Banister, and Bowling 2012, 1320; Ziegler and Schwanen 2011, 777). The same attitudes have also been described for members of immigrant communities in the United States (Lovejoy and Handy 2011, 255).

However, it is important to understand both the inclusionary *and* exclusionary effects of social networks, and that the networked society is a society of inclusions and exclusions at the same time. The exclusionary effects of social networks and the resulting mobility disadvantage will be more thoroughly investigated in Chapter 2.

Motility as a key element: individual capabilities and preferences

In order to identify unrealised mobility needs and estimate the gap between actual travel and latent mobility needs, the concept of *motility* can be employed (Figure 1.1). "Motility can be defined as how an individual or group

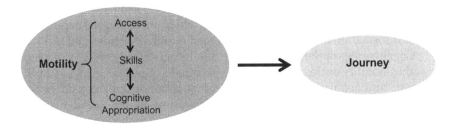

Figure 1.1 The motility approach.
Source: Authors, adapted from De Witte and Macharis (2010), based on Flamm and
Kaufmann 2006; Kaufmann, Bergman, and Joye 2004.

takes possession of the realm of possibilities for mobility and builds on it
to develop personal projects" (Flamm and Kaufmann 2006, 168). Furthermore, motility can be described "as the way in which entities [persons] access and appropriate the capacity for socio-spatial mobility according to
their circumstances" (Kaufmann, Bergman, and Joye 2004, 750).

Motility hence analyses the *potential* of and *capacity* for movement. The
study of potential movement reveals further insights into people's mobility
as well as into its wider spatial and social consequences (Kaufmann, Bergman, and Joye 2004, 749). The motility approach allows us to grasp a better
understanding of the *contextuality* of mobility challenges. It also allows us
to analyse and explain how increased travel options do not result in more
freedom and mobility for all; this lets us understand that individuals use
these options in different ways. In other words, the motility approach provides a useful concept to analyse mobility poverty empirically.

Kaufmann and his team identified three interrelated groups of factors
that define the potential to be mobile:

- **Access**: This describes the range of possible mobilities according to
  place, time and other contextual constraints. Access varies according
  to the options that are available and the *conditions* under which these
  options can be used. The options entail the available means of transportation and communication, as well as the range of services and goods
  available at a given time. The conditions refer to constraints in accessibility of the *options*, e.g. distance, cost, need to carry heavy loads.
  Spatial distribution of people and infrastructure, spatial and transport
  policies and the socio-economic position of individuals, households
  and groups are paramount to analyse access.
- **Skills**: This describes capabilities and competencies required in order
  to use mobility options. This includes acquired knowledge and organisational capacity in order to plan activities. Three aspects are central: physical ability to move from one place to another under given

circumstances; acquired skills that relate to the rules and regulations of movement (e.g. driving licence and parking permits); and organisational skills to plan and coordinate activities, including the acquisition of information and the above mentioned required abilities and skills.

- **Cognitive appropriation**: This describes the personal evaluation of available mobility options in relation to personal aspirations, plans and projects and acquired skills. This aspect most importantly considers *how* and *why* people make mobility decisions – how people consider certain options, deem them more or less appropriate for themselves and ultimately select specific options. It also considers how people evaluate their own skills and decisions. How and why people make use of available options (or not) has to do with personal needs, projects, personal aspirations, plans and so on. These needs and aspirations are interrelated with prior experiences, personal values, norms, habits, attitudes and strategies (Flamm and Kaufmann 2006, 169; Kaufmann 2011, 41–44; Kaufmann, Bergman, and Joye 2004, 750).

In studying mobility needs with the motility approach, it is possible to identify the deeper-lying elements that influence mobility or immobility behaviour, originating from various parameters. With the increasing disappearance of generally accepted organising principles and the heterogeneity of norms and values in recent times of transformation, it is important to consider how people make sense of this heterogeneity and put it into mobility practice.

Scholars in the field of mobility and transport have subsequently widened their analysis of mobility needs and patterns by including the *potential* for movement. Cresswell and Uteng for example point out that "by mobility we mean not only geographic movement but also the potential for undertaking movements (motility) as it is lived and experienced – movement and motility plus meaning plus power" (Cresswell and Uteng 2008, 2). Canzler, Kaufmann and Kesselring define mobility

> as a *change* of condition by targeting three dimensions: movements, networks and motility. [...] Movements refer to strictly a geographic dimension. [...] Networks can be defined as the framework of movements; [they] delineate the field of conceptualized possibilities. [...] Motility is how an individual or groups endorses the field of movement possibilities and uses them.
>
> (Canzler, Kaufmann, and Kesselring 2008,
> 2–3, emphasis in original)

The role of socialisation

Socialisation is one of the important formative processes that shape people's attitudes and behaviours. How individuals are integrated in society over different stages of their lives influences individuals' learning experience and formation of social roles. Personal experiences and social norms are internalised and developed into personal norms that ultimately guide behaviour. Factors typically significant for socialisation are family members, especially parents, friends, peers, colleagues, but also institutions such as schools and the media. Socialisation can be defined as "the adoption of a group's (typical) behaviours, opinions and values by an individual so that thus an individual capable of social acting emerges" (Tully and Baier 2011, 195; cited in Scheiner 2017, 392).

Socialisation is naturally also a key factor in shaping individuals' mobility needs and routines. Such processes influence the travel mode choices of people and impact on how they adapt their mobility behaviour to changing external circumstances. Research on socialisation shows that mobility behaviour is impacted at an early age by primary socialisation relating to parental and family mobility and secondary socialisation in later life by mobility education in school, by mobility behaviour of peers and cliques in adolescence and by partners (see e.g. Kroesen 2015, 492–493, 501–502). Hence, *socialisation processes are most formative in childhood and adolescence, but they are not limited to this.*

Moving to our field of analysis, travel behaviour can change over the whole life course, although changes are slower in later life (Scheiner 2017, 393). Focusing on automobility, studies show that "pro car" attitudes in car-owning households are transferred to the children, who themselves develop positive attitudes towards cars (on this issue, see also Chapter 9 of this volume). Other research suggests that the media reinforces a desire for car ownership and usage as children embrace knowledge of and desire for particular types of cars and their associated lifestyles, though the media is not the main or sole cause of how children's travel attitudes and choices develop (Baslington 2008, 109). For teenagers, having access to different mobility options is crucial for independence from family support. Thus, having experienced and being familiar with different forms of transport can enlarge the activity space and help foster social relationships (Tully and Baier 2011, 195–198).

Conversely, reduced or highly limited exposure to transport facilities creates barriers to access those mobility systems and thus leads to reduced mobility in young age (and later to lower perceived transport needs). This environment drives constraints in mobility, which can significantly impede access to education, job opportunities, leisure and social opportunities. On a different angle, *travel patterns are characterised by routines and habits* and this can lead to transport mode "decisions" *which may NOT follow "rational" choice and decision-making for the best available option.* This seems trivial, but too often it is not part of a policy maker's mindset.

We can thus state that observed travel behaviour and travel patterns can differ from the real travel needs of people. Early socialisation with cars can contribute to forced car ownership because other available options that are cheaper may be out of sight for individuals.

In a wider view, every travel behaviour is the outcome of a long process of socialisation. Also, travel needs are – actually – an outcome of how we define ourselves in the large social frame and our ability to use transport systems. Our capacity to drive a car, to know which bus to use or to manage a ride on a train are part of our travel choice and limitation.

If we tackle mobility under this angle, we therefore need to understand how we develop our mobility needs and how we are exposed and socialised to transport systems.

Travel know-how and spatial knowledge

Flamm and Kaufmann (2006, 175–176) point out that, in order to use and master means of transportation, it is crucial to acquire driving and riding know-how for any type of vehicle. Reaching a certain level of know-how is a process of accumulating experience that requires a medium- to long-term learning process. Without mastering means of transportation, individual mobility is severely restricted or impeded.

This may be obvious and most important for individual forms of transport such as car-driving and bicycle usage. Studies show that young car drivers need at least 3,000 kilometres to gain minimal experience of driving an automobile (Pervanchon 1999, 22–24, 83, in Flamm and Kaufmann 2006, 175). However, also the use of collective modes of transport demand experience and sometimes a good understanding and know-how of a certain transport regime so as to make travel possible, convenient and comfortable (Figure 1.2).

While travel experience itself is important, it is even more crucial that a person is also willing to learn from travel experiences. When a positive opinion on a certain transport mode pre-exists, learning can take place and know-how is accumulated. However, *when there is already a negative attitude towards a transport mode, these attitudes are most likely to be confirmed and improvement of the aptitude not likely* (Flamm and Kaufmann 2006, 176).

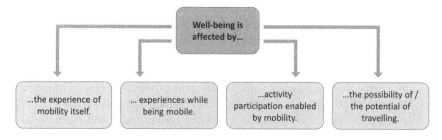

The relationships between well-being and mobility

Well-being is affected by...

| ...the experience of mobility itself. | ... experiences while being mobile. | ...activity participation enabled by mobility. | ...the possibility of / the potential of travelling. |

Figure 1.2 The relationships between well-being and mobility.
Source: Authors, adapted from Ferreira et al. 2014, based on Vos et al. 2013.

We know that when users are given a hypothetical choice between a car and public transport travel for daily commuting, car travel is associated with a more satisfactory travel experience and general well-being (Vos et al. 2013, 428).

The picture changes when actual commuting experiences are evaluated. Research results in the United States and Europe show that commuting experiences in public transport – especially train travel – are evaluated more positively than car commuting. In the US case, while commuting, car commuters experienced more stress, anxiety, impatience and less enjoyment than commuters by public transportation (Abou Zeid 2009, 83–87). A study in different European cities revealed that train and metro commuters are more satisfied with the commuting experience than car commuters; however, in this study, bus commuters were the least satisfied (Duarte et al. 2010, 22–23). This means that car usage, e.g. for commuting, is reported positively, although the actual experience may be less satisfactory. It is also important to note the gender difference in gaining travel experience. Even if access to cars and possession of a driving licence is granted, women in partnerships tend to drive less because their male partners drive when they travel together (Ryan, Wretstrand, and Schmidt 2015, 112) or the male partner is the primary user of the vehicle in single-car households (Hjorthol 2008, 206). *Not having enough experience and perceived insecurity in traffic may be the reasons why women more often than men give up driving*; in addition, women stop driving earlier than men while still able to drive (Hjorthol 2013, 1197, 1205).

As stated earlier, riding public transportation also needs to be learned. Passengers in transit are exposed to "the everyday challenges of contemporary urban living and the thrown-togetherness of different bodies" that "can solidify prejudices and antagonisms as much as it can weaken them" (Wilson 2011, 646).

In order to be able to move across space in different modes of transport, knowledge of the area travelled and of the destination are both useful. The degree of spatial competence and familiarity depends on an individual's cognitive map that s/he draws in regard to the environment. "Cognitive mapping is the process of encoding, storing, and manipulating experienced and sensed information that can be spatially referenced [...]. Parts of it are needed to solve problems, including decision-making and choice related to travel behaviour" (Golledge and Gärling 2004, 503). This knowledge is influenced by spatial thinking and reasoning; the scope and precision of these cognitive maps may be very different, thus "spatial representations in humans are incomplete and error prone" (Golledge and Gärling 2004, 506). Prior to the existence of navigation systems, spatial knowledge was essential for those using individual modes of transport or offering transport services to others. Famously, London cab drivers earlier had to pass an exam on their geographic understanding of the city to become a licensed cab driver. Apart from the knowledge gained in personal experiences, assistant tools were limited to maps and personal recommendations of others.

Spatial knowledge is essential for way-finding and successfully reaching destinations. It makes people aware of the time and cost involved. Beyond that, spatial cognition is relevant for the soft factors of travel experiences such as reliability, regularity and comfort. Kevin Lynch analysed how cities are experienced emotionally and thus differently from person to person: naturally, the personal and subjective evaluation of these experiences shapes access to a city's opportunities (Lynch 1960).

This becomes particularly important for perceived levels of safety in traffic and transport in different geographical areas. Incomplete spatial knowledge can lead to negative experiences in transport or can suppress travel needs overall. Kevin Lynch and others argue that it is actually not the knowledge about the cartographic, Euclid space that shapes preferences for movement, but the "sense of place" associated with meaning and characterised by heterogeneity (see e.g. Massey 1994). In Lynch's work, landmarks are such places that are fused with meaning and provide assistance to way-finding. In fact, in parts of the world where detailed, micro-level cartographic information is absent, way-finding instructions usually work via the indication of local landmarks. In navigation and ICT supported systems, such place-based measures are increasingly taken up in experiential and gamification approaches (see e.g. Meurer et al. 2018; Papangelis et al. 2017; Souza e Silva 2017).

The earlier issues can be translated to practical cases and demonstrated, for example, for pedestrians and cyclists. For cyclists and pedestrians, knowledge about the suitability of the cycling infrastructure and coherence of a bicycle network make an important difference in the decision for or against bicycle use. As the benefits of cycling on several levels are regaining attention, city administrations and advocacy groups are circulating more information on these aspects. However, much of the knowledge is acquired by personal experience. For cycling, personal safety is tantamount. Thus, knowledge about accident-prone areas and places is required to deal effectively with safety hazards. Manton et al. show that in this respect the perceived risks can overshadow the actual risks. A focus on perceived risks highlights how gender and cycling experience take effect on different preferences in bicycle usage and also barriers to bicycling for population groups such as the elderly (Manton et al. 2016, 19–20).

Considering the rise of hybrid transport regimes (neither public nor private such as car-pooling), it can be stated that being a *car driver* increases spatial knowledge more than being a *car passenger*. If needed at all, the responsibility for spatial knowledge is left to the driver. With current rising demand for ride-sharing services, spatial cognitive experience and thus opportunities to acquire knowledge are shrinking. It is likely that children's early socialisation with the car and parents' chauffeuring contribute to children not developing a sense of space at all. Interestingly, prior experience in long-distance travelling, e.g. for tourism purposes, together with exposure to cartographic material contributed positively to the spatial cognition of young children on the macro scale, exhibited by being able to draw a world map (Schmeinck and Thurston 2011, 10–13). In highly mobile societies, where high mobility is associated with status, such an early socialisation with the global scale may be an important element for the development of network capital (Frändberg 2009, 652–653, 663–665).

Realised and unrealised mobility: virtual life and the impact on mobility

In this final subchapter, we turn to "virtual mobilities" to illustrate the conjunction between mobility and immobility on the one hand and the gap between realised and unrealised mobility on the other hand. It will become clear that ICT has completely overthrown former certainties regarding needs, desires and motivations to move or not. However, it will also be shown that, even in times of ICT, motility – the potential to move – is dependent on access, personal aptitudes and skills regarding the use of ICT. Hence the proliferation of ICT in the sphere of mobility results in a complex picture that poses great potentials to alleviate mobility poverty and at the same time adds new challenges to the phenomenon.

Information and communications technology (ICT) has greatly changed contemporary life. ICT tools support the ease of movement through space and virtual mobility has been highlighted as a means to reduce and replace physical mobility. However, virtual proximity has only partially replaced the need for co-presence and the need for corporeal mobility. Face-to-face interaction is still important in the digital and virtual age as it fosters friendship, intimacy and trust: "As communication increases, social networks become dense and provide more and more necessity for face-to-face-meetings. Virtual activities stimulate real activities and interaction" (Kesselring and Vogl 2016, 148).

With recent developments of digitisation and augmentation of communication tools, the range of ways and modes to respond to personal needs has extended substantially. Social relations and networks, but also areas that touch people's basic needs are under deep transformation, most substantially the fields of work, education, health care, supply, access to public services and political participation.

Although these forms have not been taken up equally by all social groups in all geographic areas, there is no doubt that usage and coverage will further increase in the future. It is remarkable how ICT has changed mobility patterns (impacting both the concepts and the practice of transport poverty) and how the diffusion of digitisation in all aspects of life have led to the emergence of new mobility needs, but also raise new questions about individuals' ability to move (or stay).

Research on the impact of ICT on travel behaviour has highlighted at least three effects of ICT on mobility: (i) *modification,* (ii) *substitution* and (iii) *enhancement* or *acceleration* (see e.g. Konrad and Wittowsky 2017, 2).

There is no clear and unambiguous picture on the question of whether virtual mobility generally reduces, maintains or increases physical mobility today and in the future. The three effects of ICT are not mutually exclusive; even on the individual level, it cannot be clearly assessed whether a person is moving more or less due to ICT (Mokhtarian, Salomon, and Handy 2006, 278). As regards the benefits of ICT, it is often generally argued that digital tools in combination with mobile communication technologies can increase activity while travelling and decrease resistance to physical movement. ICT usage allows us to make travel time productive or more attractive. This effect of ICT on mobility has been called the *modifying* dimension of ICT (Tully and Alfaraz 2017, 11). Using ICT and satellite navigation reduces not only travel times but also travel time uncertainties, discomfort and the need to plan in advance (Ben-Elia and Avineri 2015, 370; van Wee 2016, 10–11).

What is more important for the analysis of mobility poverty is that, for those familiar with these technologies, *the burden of physical movement can potentially be diminished and this familiarity may even alleviate some of the disadvantages that social groups experience while being mobile.* For example, this may be the case for physically impaired people: due to real-time and location-based information systems on barrier-free facilities, travel is becoming easier or is indeed made possible in the first place. However, for those with low digital aptitude, these technologies and services are out of reach. As usage is becoming more widespread and the norm, people with less digital aptitude face challenges. Many researchers argue that virtual mobility decreases the need for physical mobility, thus it *substitutes* travel. E-shopping, e-learning and teleworking can replace the need for physical presence and hence reduce travel. In social relationships, ICT tools such as messaging and internet telephony can create a sense of proximity between people who are physically divided and thus decrease the need for physical meetings and travel (Konrad and Wittowsky 2017, 2). The relationship between mobility and immobility is therefore recalibrated. A person's motility becomes more strongly associated with a person's ability to navigate ICT systems.

Besides access to ICTs, there is another aspect that needs to be highlighted: the growing need for co-presence despite increased telework. This need for co-presence can increase the burden of mobility.

Tele- and homeworking respond to individualised and complex arrangements and are likely to increase. There is a trend of entrepreneurial co-living in Europe, especially in Scandinavian countries, where entrepreneurs live, work and socialise under the same roof (Rogel 2013; Valva 2014); such living and working arrangements reduce the need for travel and require robust digital infrastructure and uninterrupted connectivity. On the other hand, in many respects, *virtual mobility produces more travel and thus accelerates or enhances mobility.* As already pointed out, social relations and networks can be maintained via a wide array of communication tools. However, in order to maintain and secure relationships, moments of co-presence are more important than ever.

Thus, with growing networks and distances, *the need for physical mobility to nurture these networks and fulfil social obligations is also increasing.* Elliott and Urry have highlighted the changing nature of tourist-type travel in this regard. Visiting friends and relatives involving middle and long-distance travelling has become a substantial part of leisure travel (Elliott and Urry 2010, 53–57). As touched upon earlier, despite growing ICT penetration, or precisely because of that, there is evidence that business travel is likely to increase and not decrease. A study in France showed that high trip frequency and demand in business travel above 80 km is no longer restricted to persons with a high income and work responsibility, such as executives, but also intermediary professionals (Aguiléra and Proulhac 2015, 34). This supports the observation that long-distance travel is increasingly becoming a prerequisite in contemporary employment and disadvantages those who are not able to conduct physical travel frequently. The aspect of mobility burden will be investigated more thoroughly in Chapter 2.

References

Abou Zeid, Maya. 2009. "Measuring and modeling activity and travel well-being." Massachusetts Institute of Technology. https://its.mit.edu/sites/default/files/documents/PhD%20Thesis_Maya%20Abou-Zeid.pdf, accessed on 19 March 2020.

Aguiléra, Anne, and Laurent Proulhac. 2015. "Socio-occupational and geographical determinants of the frequency of long-distance business travel in France." In *Journal of Transport Geography* 43: 28–35. https://doi.org/10.1016/j.jtrangeo.2015.01.004.

Baslington, Hazel. 2008. "Travel socialization: A social theory of travel mode behavior." In *International Journal of Sustainable Transportation* 2 (2): 91–114. https://doi.org/10.1080/15568310601187193.

Ben-Elia, Eran, and Erel Avineri. 2015. "Response to travel information: A behavioural review." In *Transport Reviews* 35 (3): 352–377. https://doi:10.1080/01441647.2015.1015471.

Canzler, Weert, Vincent Kaufmann, and Sven Kesselring. 2008. "Tracing mobilities: An introduction." In *Tracing Mobilities: Towards a Cosmopolitan Perspective*, edited by Weert Canzler, Vincent Kaufmann and Sven Kesselring, 1–10. Aldershot: Ashgate.

Cass, Noel, Elizabeth Shove, and John Urry. 2005. "Social exclusion, mobility and access." In *The Sociological Review* 53 (3): 539–555. https://doi.org/10.1111/j.1467-954X.2005.00565.x.

Cresswell, Tim, and Tanu P. Uteng. 2008. "Gendered mobilities: Towards an holistic understanding." In *Gendered Mobilities,* edited by Tanu P. Uteng and Tim Cresswell, 1–12. Aldershot: Ashgate.

De Witte, Astrid, and Cathy Macharis. 2010. "Commuting to Brussels: How attractive is "free" public transport?" In *Brussels Studies.* Collection générale, n° 37. https://doi.org/10.4000/brussels.755

Duarte, André, Camila Garcia, Grigoris Giannarakis, Susana Limão, Amalia Polydoropoulou, and Nikolaos Litinas. 2010. "New approaches in transportation planning: Happiness and transport economics." In *NETNOMICS: Economic Research and Electronic Networking* 11 (1): 5–32.

Elliott, Anthony, and John Urry. 2010. *Mobile Lives.* New York: Routledge.

Ferreira, Antonio, Luca Bertolini, Petter Næss, and Greg Marsden. 2014. *Immobility as well-being. Creating alternatives to pro-mobility discourses.* Presentation at the Royal Geographical Society (RGS) Annual International Conference, 27–29 August 2014. https://www.slideshare.net/ITSLeeds/immobility-as-wellbeing creating-alternatives-to-promobility-discourses, accessed 10 March 2018

Flamm, Michael, and Vincent Kaufmann. 2006. "Operationalising the concept of motility: A qualitative study." In *Mobilities* 1 (2): 167–189. https://doi.org/10.1080/17450100600726563.

Frändberg, Lotta. 2009. "How normal is travelling abroad? Differences in transnational mobility between groups of young Swedes." In *Environment and Planning A* 41 (3): 649–667. https://doi.org/10.1068/a40234.

Golledge, Reginald G., and Tommy Gärling. 2004. "Cognitive maps and urban travel." In *Handbook of Transport Geography and Spatial Systems*, 501–512: Emerald Group Publishing Limited. https://doi.org/10.1108/9781615832538-028.

Hjorthol, Randi. 2008. "Daily mobility of men and women – A barometer of gender equality?" In *Gendered Mobilities,* edited by Tanu P. Uteng and Tim Cresswell, 193–210. Aldershot: Ashgate.

Hjorthol, Randi. 2013. "Transport resources, mobility and unmet transport needs in old age." In *Ageing & Society* 33: 1190–1211. https://doi.org/10.1017/S0144686X12000517.

Jansuwan, Sarawut, Keith M. Christensen, and Anthony Chen. 2013. "Assessing the transportation needs of low-mobility individuals: Case study of a small urban community in Utah." In *Journal of Urban Planning and Development* 139 (2): 104–114. https://doi.org/10.1061/(ASCE)UP.1943-5444.0000142.

Kaufmann, Vincent. 2011. *Rethinking the City: Urban Dynamics and Motility.* Lausanne: EPFL Press.

Kaufmann, Vincent, Manfred M. Bergman, and Dominique Joye. 2004. "Motility: Mobility as capital." In *International Journal of Urban and Regional Research* 28 (4): 745–756. https://doi.org/10.1111/j.0309-1317.2004.00549.x.

Kesselring, Sven, and Gerlinde Vogl. 2016. "'…Travelling, where the opponents are': Business travel and the social impacts of the new mobilities regimes." In *International Business Travel in the Global Economy*, edited by Ben Derudder and Frank Witlox, 145–162. London and New York: Routledge.

Konrad, Kathrin, and Dirk Wittowsky. 2017. "Virtual mobility and travel behavior of young people – Connections of two dimensions of mobility." In *Research in Transportation Economics* 68: 11–17. https://doi:10.1016/j.retrec.2017.11.002.

Kroesen, Maarten. 2015. "Do partners influence each other's travel patterns? A new approach to study the role of social norms." In *Transportation Research Part A: Policy and Practice* 78: 489–505. https://doi.org/10.1016/j.tra.2015.06.015.

Lovejoy, Kristin, and Susan Handy. 2011. "Social networks as a source of private-vehicle transportation: The practice of getting rides and borrowing vehicles among Mexican immigrants in California." In *Transportation Research Part A: Policy and Practice* 45 (4): 248–257. https://doi.org/10.1016/j.tra.2011.01.007.

Lucas, Karen. 2006. "Providing transport for social inclusion within a framework for environmental justice in the UK." In *Transportation Research Part A: Policy and Practice* 40 (10): 801–809. https://doi.org/10.1016/j.tra.2005.12.005.

Lynch, Kevin. 1960. *The Image of the City*. Cambridge, MA and London: MIT Press.

Manton, Richard, Henrike Rau, Frances Fahy, Jerome Sheahan, and Eoghan Clifford. 2016. "Using mental mapping to unpack perceived cycling risk." In *Accident; Analysis and Prevention* 88: 138–149. https://doi.org/10.1016/j.aap.2015.12.017.

Massey, Doreen. 1994. *Space, Place, and Gender*. Minneapolis: University of Minnesota Press.

Meurer, Johanna, Martin Stein, David Randall, and Volker Wulf. 2018. "Designing for way-finding practices – A study about elderly people's mobility." In *International Journal of Human-Computer Studies* 115: 40–51. https://doi.org/10.1016/j.ijhcs.2018.01.008.

Mokhtarian, Patricia L., Ilan Salomon, and Susan L. Handy. 2006. "The impacts of ICT on leisure activities and travel: A conceptual exploration." In *Transportation* 33 (3): 263–289. https://doi.org/10.1007/s11116-005-2305-6.

Nordbakke, Susanne, and Tim Schwanen. 2015. "Transport, unmet activity needs and wellbeing in later life: Exploring the links." In *Transportation* 42 (6): 1129–1151. https://doi.org/10.1007/s11116-014-9558-x.

Papangelis, Konstantinos, Melvin Metzger, Yiyeng Sheng, Hai-Ning Liang, Alan Chamberlain, and Ting Cao. 2017. "Conquering the city: Understanding perceptions of mobility and human territoriality in location-based mobile games." In *Proceedings of the ACM on Interactive, Mobile, Wearable and Ubiquitous Technologies* 1 (3): 90. https://doi.org/10.1145/3130955.

Pereira, Rafael H. M., Tim Schwanen, and David Banister. 2017. "Distributive justice and equity in transportation." In *Transport Reviews* 37 (2): 170–191. https://doi.org/10.1080/01441647.2016.1257660.

Pyer, Michelle, and Faith Tucker. 2014. "'With us, we, like, physically can't': Transport, mobility and the leisure experiences of teenage wheelchair users." In *Mobilities* 12 (1): 36–52. https://doi.org/10.1080/17450101.2014.970390.

Rajé, Fiona. 2007. "The lived experience of transport structure: An exploration of transport's role in people's lives." In *Mobilities* 2 (1): 51–74. https://doi.org/10.1080/17450100601106260.

Rittner, Barbara, and Alan B. Kirk. 1995. "Health care and public transportation use by poor and frail elderly people." In *Social Work* 40 (3): 365–373. https://doi.org/10.1093/sw/40.3.365.

Rogel, Liat. 2013. "HousingLab: A laboratory for collaborative innovation in urban housing." *Politecnico di Milano*. https://www.politesi.polimi.it/bitstream/10589/82782/1/Rogel_Liat_Dissertation_s.pdf, accessed on 19 March 2020.

Ryan, Jean, Anders Wretstrand, and Steven M. Schmidt. 2015. "Exploring public transport as an element of older persons' mobility: A capability approach perspective." In *Journal of Transport Geography* 48: 105–114. https://doi.org/10.1016/j.jtrangeo.2015.08.016.

Scheiner, Joachim. 2017. "Mobility biographies and mobility socialisation—New approaches to an old research field." In *Life-Oriented Behavioral Research for Urban Policy*, edited by Junyi Zhang, 385–402. Tokyo: Springer Japan.

Schmeinck, Daniela, and Allen Thurston. 2011. "The influence of travel experiences and exposure to cartographic media on the ability of ten-year-old children to draw cognitive maps of the world." In *Scottish Geographical Journal* 123 (1): 1–15. https://doi.org/10.1080/00369220718737280.

Schwanen, Tim, David Banister, and Ann Bowling. 2012. "Independence and mobility in later life." In *Geoforum* 43 (6): 1313–1322. https://doi.org/10.1016/j.geoforum.2012.04.001.

Souza e Silva, Adriana de. 2017. "Pokémon go as an HRG: Mobility, sociability, and surveillance in hybrid spaces." In *Mobile Media & Communication* 5 (1): 20–23. https://doi.org/10.1177/2050157916676232.

Tully, Claus, and Claudio Alfaraz. 2017. "Youth and mobility: The lifestyle of the new generation as an indicator of a multi-local everyday life." In *Applied Mobilities* 2 (2): 182–198. https://doi.org/10.1080/23800127.2017.1322778.

Tully, Claus, and Dirk Baier. 2011. "Mobilitätssozialisation." In *Verkehrspolitik*, edited by Oliver Schwedes, 195–211. Wiesbaden: Springer. https://doi.org/10.1007/978-3-531-92843-2_10.

Urry, John. 2007. *Mobilities*. Cambridge: Polity.

Urry, John, and Margaret Grieco, eds. 2011. *Mobilities: New Perspectives on Transport and Society*. London and New York: Routledge.

Uteng, Tanu P., and Tim Cresswell, eds. 2008. *Gendered Mobilities*. Aldershot: Ashgate.

Valva, Paul. 2014. "Shared living and sustainability: Emerging trends in the tourism industry." In *Almatourism-Journal of Tourism, Culture and Territorial Development* 5 (3): 1–18. https://doi.org/10.6092/issn.2036-5195/4618.

van Wee, Bert. 2016. "Accessible accessibility research challenges." In *Journal of Transport Geography* 51: 9–16. https://doi.org/10.1016/j.jtrangeo.2015.10.018.

Vos, Jonas de, Tim Schwanen, Veronique van Acker, and Frank Witlox. 2013. "Travel and subjective well-being: A focus on findings, methods and future research needs." In *Transport Reviews* 33 (4): 421–442. https://doi.org/10.1080/01441647.2013.815665.

Wilson, Helen F. 2011. "Passing propinquities in the multicultural city: The everyday encounters of bus passengering." In *Environment and Planning A* 43 (3): 634–649. https://doi.org/10.1068/a43354.

Ziegler, Friederike, and Tim Schwanen. 2011. "'I like to go out to be energised by different people': An exploratory analysis of mobility and wellbeing in later life." In *Ageing & Society* 31 (5): 758–781. https://doi.org/10.1017/S0144686X10000498.

2 Unequal mobilities, network capital and mobility justice

Tobias Kuttler and Massimo Moraglio

Abstract

In this chapter, we continue our investigation of the gap between realised and unrealised mobility. After focusing on individual *mobility needs and preferences* in the previous chapter, we now turn to examining notions of *unequal mobilities* and *mobility justice* beyond the distribution of accessibility levels. Therefore, in this chapter, we try to understand mobility poverty from a systemic and structural perspective.

Two different approaches to unequal mobilities will be outlined here: first, the question of mobility justice will be discussed and, second, unequal mobilities in times of the networked society. This chapter concludes with final remarks on practical and policy implications.

From transport justice and mobility justice

Access to mobility options is highly uneven according to gender, race, class, income and age. While it is important to analyse how discriminatory practices explicitly prevent the mobility of individuals from different groups, it is also necessary to investigate the historical, spatial and cultural context of uneven mobilities. Discourses and policy debates often fail to analyse these structural aspects and therefore unintentionally overlook the mobility needs of different groups (Sheller 2015 for ethnic minorities; Uteng and Cresswell 2008 for gender aspects).

This is most evident regarding the needs of women because

> when policy makers debate mobility systems, or designers implement new technologies, or researchers study new mobilities, they are unconsciously already working within a context of deeply gendered discourses that must be brought to the foreground if we are to understand how planning decisions may be contributing to unequal mobility outcomes for men and women.
>
> (Sheller 2008, 258)

Questions of justice in mobility have been raised more thoroughly recently, most prominently by Sheller (2018) and Cook and Butz (2018). Recent debates on mobility justice are at least partially rooted in endeavours to develop a theory of transport justice and equity in transport provision. Therefore, the approach of transport justice will be briefly outlined here before discussing the contours of mobility justice theories.

In the transport justice debate, the most important criterion is equity in accessibility levels. Martens states: "[A] transportation system is fair if, and only if, it provides a sufficient level of accessibility to all under most circumstances" (Martens 2017, 151). From a theoretical standpoint, there are different approaches to "fairness" in transport that define what is a sufficient level of accessibility. These different approaches are variously helpful to understand transport injustice.

In an **egalitarian** approach to transport justice, the moral guiding principle is that everyone should obtain the same level of service and access. It considers fairness as a matter of relative distribution of benefits and burdens in transportation and focuses on inequality between social groups or geographical areas. Thus, the egalitarian approach asks why certain social groups or geographic spaces and regions have higher or lower accessibility levels, or more or less transport services than others (Pereira, Schwanen, and Banister 2017, 178).

A **sufficientarian** approach is more directly concerned with transport disadvantages and meeting the basic needs of social groups vulnerable to exclusion. The moral guiding principle "is not that everyone should have the same, but that each should have enough. If everyone had enough, it would be of no moral consequence whether one had more than others" (Frankfurt 1987, 21). Thus, a sufficiency approach in transport provision aims at the avoidance of misery that is experienced under certain thresholds. Accordingly, interventions in transport systems need to prioritise benefiting people below the threshold compared to benefiting people above the threshold (Martens, Di Ciommo, and Papanikolaou 2014, 7).

These two first approaches may be in conflict with each other due to their nature. Furthermore, it is difficult to apply a strict threshold of accessibility levels because the definition of what levels of individuals' activity participation is "reasonable" and "normal" is highly relational and socially, temporally and geographically context-specific.

Thus, a third, somewhat conciliating approach has been proposed, called **prioritarianism**. Here, the proponents' point of view is that benefits matter and that they matter more the worse off the person is to whom the benefits accrue. In this approach, the moral value of a benefit, or the disvalue of a burden, diminishes as its recipient becomes better off (Casal 2007). Regarding the priority of intervention, such an approach suggests a weighing of benefits that depends on the position of a person in the range of accessibility and service levels. In practical terms, this implies a ranking of population groups according to accessibility levels. The value of the accessibility benefits diminishes the higher a person's accessibility level already is (Martens, Di Ciommo, and Papanikolaou 2014, 8).

Fourthly, "Capability Approaches" to transport justice have been highlighted, based on the works by Amartya Sen and Martha Nussbaum. In the context of these Capability Approaches, the focus of transport policies should be on guaranteeing all individuals a minimum level of access to key activities that allow basic needs (commerce, education, healthcare, employment) to be met. It should not mean, however, guaranteeing that all people have access to the same transport conditions. This approach faces two challenges: first, the identification of the minimum acceptable thresholds of accessibility, dependent on a given society's values and history and requiring politically democratic, legitimate decisions; and, second, the fact that accessibility is the result of a combination of personal characteristics and social, economic and environmental specifications. The Capability Approach therefore needs accessibility to be addressed as an attribute of individuals (and personal characteristics such as gender, age, social class, disabilities, time and income) interacting with their environment. Many authors rather implicitly follow the Capability Approach, aiming at determining the minimum level of accessibility that a transport system must offer to each vulnerable segment of the population. Yet, not all authors follow this approach explicitly, leading to different interpretations as to how the transport system should evolve (Pereira, Schwanen, and Banister 2017, 178).

More recently, there has been a significant advance to highlight the limits of the transport justice debate. Proponents of mobility justice argue that the approach of transport justice is too limited to understand the full picture of mobility disadvantage. Sheller argues that "[i]ncreasing access to transport […] will not solve the problem if we ignore the underlying processes and relations that produce mobility injustice, and which tunnel beneath transport (into the body) and beyond the city (into the world)" (Sheller 2018, 15). She further argues that, in order to understand how the movement of people, resources and information are controlled and governed, mobility justice needs to be addressed at different scales. More concretely, this means that mobility should not only be viewed at the meso-level in terms of everyday transportation, but also on the level of the body (micro-level) and transnational and global levels (macro-levels). Only when viewing unequal mobilities in conjunction with these scales is a theory of mobility justice comprehensive (Sheller 2018, 14). In terms of theoretical underpinnings, this means that a theory of mobility justice needs to pay explicit attention to discrimination and marginalisation along the lines of gender, class, race and caste. Such perspectives allow forms of privilege in, and exclusion from, deliberative processes that presuppose governance and control of movement, including everyday transport, to be identified. Such a theory furthermore includes spatial and "right to the city" perspectives (Sheller 2018, 22–32).

These perspectives follow the understanding that space cannot be containerised into different separated entities such as "cities" or "nations" or other forms of territories. Spaces are rather highly unequal and contested and therefore political, and the production of social space – in the sense of Henri Lefebvre – is a highly uneven process. With an approach that is sensitive to the historical, deeply entrenched injustices on the one hand and an understanding of the politics of space and scales on the other, an import step is taken towards envisioning actual freedom to be im/mobile.

Consequently, proponents of mobility justice do not only call for distributive justice; they argue that mobility justice can only be achieved when the power imbalances in access to information, participation and decision-making processes (deliberative and procedural justice), the recognition of historical injustice and oppressions (restorative justice) and a shift in hitherto practised forms of production of knowledge (epistemic justice) are addressed (Cook and Butz 2018, 5–19; Sheller 2018, 30–35).

For our approach to mobility poverty, as developed in this volume, the understanding of mobility justice is crucial. Only when the deeply engrained inequalities and injustices are taken into account can the gap between realised and unrealised mobility be identified. Only then do unarticulated needs, desires and motivations for mobility become uncovered and can be put to debate. Mobility injustice on all scales will be examined throughout this volume in a thorough examination of spatial aspects (Chapters 4–6) and social criteria (Chapters 7–12), both in theory and in the field (Chapters 13–18).

To open up the view even more, the next subchapter will provide further scrutiny of the political economy of differentiated mobilities.

Unequal mobilities in the networked society and the burden of mobility

The process of contemporary globalisation and economic competitiveness increase the complexity of mobility regimes, including their power relations. It is important to highlight the close ties between communication tools and mobility options because such an interplay allows more effective organisation of everyday life as well as social and business relationships. ICT drives further the extension and differentiation of social networks. It also allows schedules that are ever more complex and individualised living arrangements. Under such framing conditions, in the modern age, power relations in societies are increasingly building on the realised levels of communication and movement, the distinctive factor being the ability to "keep up" with technological innovation and social trends (Hannam, Sheller, and Urry 2006, 12; Shove 2002, 4). It has been argued that these developments further benefit those who already enjoy privileges, while those facing disadvantages may experience additional burdens (Elliott and Urry 2010, 59).

Furthermore, travelling has become a marker for status among young people, slowly replacing other status symbols such as the car (Canzler and Knie 2016, 61). Whenever new technologies emerge, the potential and opportunities to use these technologies change and those innovations may only be accessible to certain sections of the population because of high costs, the expert knowledge involved and so on (Cass, Shove, and Urry 2005, 542).

As outlined previously, the range of mobility options is further increasing day by day and so is the potential burden of mobility. In the transformation towards digital and automated societies, everyone's mobility arrangements are – in one way or another – influenced by the friction between entrenched norms and roles and the fast pace of contemporary life. While for many, negotiating these complexities has become part of everyday life, it can be argued that those who face a social or mobility-related disadvantage may be overly burdened by coping with the increased necessity of being mobile.

A key concept for understanding differentiated mobilities from a systemic perspective is the concept of "network capital" developed by John Urry. He and his colleagues argue that in the modern mobile society, above all other, it is movement and its related opportunities that have become associated with the understanding of a "good life". Social status, recognition and prestige are gained, maintained and enhanced by a person's degree of personal mobility and associated mobile lifestyles (Urry 2007).

Drawing on Pierre Bourdieu's works, it is argued that mobile lifestyles together with economic, social and cultural capital produce symbolic power that is the prime currency of social distinction and the mechanism for social stratification in contemporary life. Urry coined the term "network capital" to describe the elements needed to gain such power. He defines it as "the capacity to engender and sustain social relations with those people who are not necessarily proximate, which generates emotional, financial and practical benefit" (Elliott and Urry 2010, 59). For an in-depth understanding of mobility poverty, it is necessary to understand how network capital varies between social groups and how accumulation of such capital creates social inequalities.

The degree of network capital a person possesses depends on the degree of access to the following core elements and capabilities (Urry 2007):

- Appropriate documents, passports, visas, money, vaccines, data-readiness, qualifications and so on that enable safe movement from one place to another;
- A capability to connect with others (workmates, friends and family members) at a distance;
- Movement capacities in relationship to the environment – including physical abilities, competencies to access (digital) information and organisational skills;
- Location-free information and contact points; communication devices and mobile data access;
- Appropriate, safe and secure meeting places; and
- Access to technical systems including: cars, road space, fuel, lifts, aircraft, trains, bikes, phones, email and time and other resources to manage all of these, especially when there is a system failure.

As opportunities for travel and communication increase day by day, mobile lifestyles are not limited to the kinetic elites (Sheller 2018), but have become prevalent in the middle classes. Lifestyles of high mobility are increasingly perceived as the norm and not the exception in modern societies. High mobility or "hypermobility is glamorized, [...] idealized and made desirable in the contemporary world" (Cohen and Gössling 2015, 1667).

Several authors have argued that social media contributes to the new role of long-distance travel, especially using aviation, as a generator of social status among young people. Social media enables a constant comparison of travel patterns, the estimation of personal "travelness" and thus contributes to identity formation and self-construction (Gössling 2017, 163–164; Gössling and Stavrinidi 2015, 736–743). Thus, such forms of travel enable the accumulation of social and network capital at an early age.

Furthermore, the need to sustain growing networks of family, friends and weak ties across larger distance requires regular physical meetings and hence increased travel (Larsen, Urry, and Axhausen 2006, 109–110). Thus, in mobile societies, individuals may experience a state of anxiety about "being disconnected by those moving around, [...] being stuck in place, [...] being too localist and not networked enough" (Elliott and Urry 2010, 47), and therefore being assigned a lower social status.

These examples show that it is important to understand both the inclusionary and exclusionary effects of social networks, the "Janus-faced capacities of social capital" (Schwanen et al. 2015, 132). Social ties are a capital as outlined by Pierre Bourdieu, i.e. a resource which can be accumulated. This represents a power relationship and a power resource. Social capital has an exchange value and is not detached from material resources (and other forms of capital such as cultural capital, Schwanen et al. 2015, 127–128). Indeed, maintaining social ties needs an investment of material resources.

The networked society is thus a society of inclusions and exclusions at the same time. Hence, due to the linkages between network capital, material resources and other forms of capital, it must be assumed that, in the present mobile societies, travel and communication options are highly accessible to the "travel rich", those who are usually also the better off and benefit them more than the travel poor.

Thus, social groups that face social or transport-related disadvantages may also face a higher burden to realise mobility that is needed for nurturing both basic (socially constructed) needs and their social networks. These perspectives have greatly influenced the transport and mobility disciplines in the last 20 years. Often, however, this approach is still a blind spot in transport policy and planning: we should keep in mind how it has become increasingly difficult to translate empirical findings into policy measures and technological solutions.

This demands an analysis of mobility needs that takes into account how social ties are maintained (or not) and the extent of the "mobility burden" for vulnerable groups needed for establishing social capital. Such analysis needs to consider how vulnerable groups rely more strongly than others on additional assistance in meeting their mobility needs. Apart from a state's measures, this assistance is often support from family and friends.

From a policy perspective, it is argued that network capital should be enlarged and spread as equally as possible in order to lessen social exclusion. "A socially inclusive society would elaborate and extend the capabilities of co-presence to all its members. It would minimize 'coerced mobility', both to improve psychic health and to heighten equality". Transport policy and planning should therefore promote networking and people's freedom to meet each other and conduct relationships over larger distances (Elliott and Urry 2010, 64). Such perspectives stress that this is indeed necessary to increase mobility in addition to accessibility targets.

However, as authors point out, this social target is in conflict with environmental targets, as growing mobility for all presupposes huge and growing supplies of various resources and further drives global warming (Elliott and Urry 2010, ibid). Such an egalitarian demand is furthermore problematic because of an inherent paradox of contemporary mobility.

When networking and mobility are indeed marking social status, increasing the capabilities for movement may rather not create equality in society, but foster individual advancement and a competition for status that is by nature not egalitarian. Kaufmann concludes that "contemporary forms of mobility [...] are as much a factor of inequality as of equality: they constitute a resource that is inequitably distributed within society, while fostering access to other resources inequitably distribute in space" (Kaufmann and Montulet 2008, 54).

Final remarks: practical and policy implications

As highlighted at the beginning of this chapter, much of the literature on transport justice focuses on accessibility and accessibility poverty. It is widely acknowledged that accessibility is necessary to expand people's freedom of choice and it promotes equality of opportunity regarding employment, health care, education and other basic needs and services. Thus, from the perspective of transport policy, the focus should be on improving access to places, activities and opportunities.

The sole focus on transport and accessibility, however, may overlook the unfulfilled mobility needs of less mobile social groups. The reason for such a blind eye is that accessibility approaches – whether egalitarian or sufficientarian – need to make assumptions either on the level of inequality that is acceptable in a fair society or on a minimum level of accessibility that should be available to everyone. In practice, however, both assumptions pose substantial difficulties. First, for an egalitarian approach, it is very difficult to judge the level of inequality that is acceptable in society and, consequently, many studies avoid making clear statements on this question (Pereira, Schwanen, and Banister 2017, 178).

Second, defining a minimum threshold for levels of accessibility is equally difficult; those trying to define such a threshold may fall into the trap of making generalised assumptions about people's needs, with the consequence that diverse preferences of vulnerable social groups may be overlooked (Preston and Rajé 2007, 159). Another weakness of approaches that only focus on transport is that improving accessibility is mostly viewed in the context of social exclusion. The above outlined challenge of defining adequate levels of accessibility is then exacerbated by weak definitions of what it means to be socially included or excluded.

With the transformation of social relationships due to recent advances in communication technology and mobility options, this challenge is further complicated. Cass et al. remind us of the changing conditions of contemporary life:

> In [...] an increasingly mobile world the challenge of accessing other people, places and services at some geographical distance is not something fixed and easily measurable. What is necessary for full 'social' inclusion varies as the means and modes of mobility change and as the potential for 'access' develops with the emergence of new technologies [...]. These developments transform what is 'necessary' for full social inclusion. It is important but very difficult to acknowledge the temporal as well as the spatial dimensions of social exclusion, as these relate to the changing spatial and temporal organization of contemporary life.
>
> (Cass, Shove, and Urry 2005, 542)

A practical example of the changing spatial and temporal organisation is the varying and flexible daily schedules of people. To coordinate different aspects of everyday life with the schedules of institutions and people is a demanding task. Only when the varying schedules of households and individuals can be brought in line with the arrangements of transport systems and the varying schedules of facilities and opportunities to be reached can people's (mobility) needs be adequately met.

This leads us to define a relational approach to accessibility in which policies of accessibility should be considered "relational" in their timing and geographies (Qviström 2015). Social networks are crucial for social identity and maintaining a meaningful life, and – in the form of social capital – for acquiring material benefits and social status (Schwanen et al. 2015, 127–128). As outlined above, for those depending on family, friends and other social informal ties as a life-support system (information on jobs, support in travelling), increased need for travel and communication can indeed be a burden.

The dynamics of social networks are difficult to approach with accessibility perspectives, as they result in very specific and not generalisable mobility needs, which differ substantially from the basic "formal" needs of life.

To summarise, it can be stated that, in the contemporary world, which is characterised by an "infinity of promised or assumed opportunities arising from movement" (Elliott and Urry 2010, 8), accessibility is a necessary, but not sufficient condition for meeting people's diverse needs.

Due to the flexibility of space-time arrangements in everyday life and the transformation of social networks, any action against mobility poverty should also consider it crucial to relieve people from any additional and unnecessary burden related to their mobility. Transport regimes should be customer-friendly, easily accessible not just in terms of physical accessibility but also in term of skills and duties needed to use the service. Due to the increased complexity of transport-related activities, the coordination of transport activities should be kept at basic levels. This applies first to multimodal journeys, but it should not be limited to them. In order to enhance mobility, we state that we need to focus on people's capabilities to access desired destinations and possibly uncover hidden needs.

References

Canzler, Weert, and Andreas Knie. 2016. "Mobility in the age of digital modernity: Why the private car is losing its significance, intermodal transport is winning and why digitalisation is the key." In *Applied Mobilities* 1 (1): 56–67. https://doi.org/10.1080/23800127.2016.1147781.

Casal, Paula. 2007. "Why sufficiency is not enough." In *Ethics* 117 (2): 296–326. https://doi.org/10.1086/510692.

Cass, Noel, Elizabeth Shove, and John Urry. 2005. "Social exclusion, mobility and access." In *The Sociological Review* 53 (3): 539–555. https://doi.org/10.1111/j.1467-954X.2005.00565.x.

Cohen, Scott A., and Stefan Gössling. 2015. "A darker side of hypermobility." In *Environment and Planning A* 47 (8): 1661–1679. https://doi.org/10.1177/0308518X15597124.

Cook, Nancy, and David Butz. 2018. *Mobilities, Mobility Justice and Social Justice.* London: Routledge.

Cresswell, Tim, and Tanu P. Uteng. 2008. "Gendered mobilities: Towards an holistic understanding." In *Gendered Mobilities*, edited by Tanu P. Uteng and Tim Cresswell, 1–12. Aldershot: Ashgate.

Elliott, Anthony, and John Urry. 2010. *Mobile Lives.* New York: Routledge.

Frankfurt, Harry. 1987. "Equality as a moral ideal." In *Ethics* 98 (1): 21–43. http://www.jstor.org/stable/2381290.

Gössling, Stefan. 2017. "ICT and transport behavior: A conceptual review." In *International Journal of Sustainable Transportation* 12 (3): 153–164. https://doi.org/10.1080/15568318.2017.1338318.

Gössling, Stefan, and Iliada Stavrinidi. 2015. "Social networking, mobilities, and the rise of liquid identities." In *Mobilities* 11 (5): 723–743. https://doi.org/10.1080/17450101.2015.1034453.

Hannam, Kevin, Mimi Sheller, and John Urry. 2006. "Mobilities, immobilities and moorings." In *Mobilities* 1 (1): 1–22. https://doi.org/10.1080/17450100500489189.

Kaufmann, Vincent, and Bertrand Montulet. 2008. "Between social and spatial mobilities: The issue of social fluidity." In *Tracing Mobilities: Towards a Cosmopolitan Perspective*, edited by Weert Canzler, Vincent Kaufmann, and Sven Kesselring, 38–55. Aldershot: Ashgate.

Larsen, Jonas, John Urry, and Kay Axhausen. 2006. *Mobilities, Networks, Geographies.* Farnham: Ashgate.

Martens, Karel. 2017. *Transport Justice.* New York and London: Routledge.

Martens, Karel, Floridea Di Ciommo, and Anestis Papanikolaou. 2014. "Incorporating equity into transport planning: Utility, priority and sufficiency approaches." In *Proceedings of the XVIII Congreso Panamericano de Ingeniería de Tránsito, Transporte y Logística*, Santander, Spain.

Pereira, Rafael H. M., Tim Schwanen, and David Banister. 2017. "Distributive justice and equity in transportation." In *Transport Reviews* 37 (2): 170–191. https://doi.org/10.1080/01441647.2016.1257660.

Preston, John, and Fiona Rajé. 2007. "Accessibility, mobility and transport-related social exclusion." In *Journal of Transport Geography* 15 (3): 151–160. https://doi.org/10.1016/j.jtrangeo.2006.05.002.

Qvistrӧm, Mattias. 2015. "Putting accessibility in place: A relational reading of accessibility in policies for transit-oriented development." In *Geoforum* 58: 166–173. https://doi.org/10.1016/j.geoforum.2014.11.007.

Schwanen, Tim, Karen Lucas, Nihan Akyelken, Diego C. Solsona, Juan-Antonio Carrasco, and Tijs Neutens. 2015. "Rethinking the links between social exclusion and transport disadvantage through the lens of social capital." In *Transportation Research Part A: Policy and Practice* 74: 123–135. https://doi.org/10.1016/j.tra.2015.02.012.

Sheller, Mimi. 2008. "Gendered mobilities: Epilogue." In In *Gendered Mobilities,* edited by Tanu P. Uteng and Tim Cresswell, 257–265. Aldershot: Ashgate.

Sheller, Mimi. 2015. "Racialized mobility transitions in Philadelphia: Connecting urban sustainability and transport justice." In *City & Society* 27 (1): 70–91. https://doi.org/10.1111/ciso.12049.

Sheller, Mimi. 2018. *Mobility Justice: The Politics of Movement in an Age of Extremes.* Brooklyn: Verso Books.

Shove, Elizabeth, ed. 2002. *Rushing Around: Coordination, Mobility and Inequality.* https://www.lancaster.ac.uk/staff/shove/choreography/rushingaround.pdf, accessed 6 May 2020.

Urry, John. 2007. *Mobilities.* Cambridge: Polity Press.

Uteng, Tanu P., and Tim Cresswell, eds. 2008. *Gendered Mobilities.* Aldershot: Ashgate.

3 The impact of life events on travel behaviour

Delphine Grandsart

Abstract

In the context of dealing with mobility poverty, it is relevant to consider how personal choices and (changes in) one's personal situation affect behaviour and, possibly, lead to behavioural change. A growing number of studies have explored and confirmed the impact of life events on travel behaviour. As most travel behaviour is habitual, it is a challenge to encourage people to consider other transport modes. For policy makers and transport service providers, it follows that life events or transition points can – and should – be considered as opportunities for triggering behavioural change, i.e. for promoting alternative transport options and nudging people towards using them.

Life events in travel behaviour change theory: an overview

Personal choices and (changes in) one's personal situation affect behaviour and may trigger behavioural change. In this chapter, we analyse how behaviour and behavioural change insights can be applied to the study of travel behaviour and travel behaviour change. First, we address a more theoretical framework and then we offer some case studies.

We can start with Adjei and Behrens (Adjei et al. 2012). In their review and synthesis of travel behaviour theories and experiments, they distinguish between four types of theories, depending on which questions they seek to answer:

- How behavioural choices are made (rational choice theory, prospect theory, habit formation theory, theory of interpersonal behaviour);
- What factors affect decision-making (theory of planned behaviour, theory of interpersonal behaviour, norm activation theory);
- When behavioural change occurs (habit formation theory, cognitive dissonance theory, stages of change model); and
- How people respond to behaviour change interventions (self-perception theory, goal setting theory).

They conclude that "Rational Choice Theory" and "Theory of Planned Behavior" have so far been dominant as the underlying framework for travel behaviour analysis and experiments, even though other approaches – notably prospect theory, habit formation theory, goal setting theory – are also receiving growing attention.

According to the theory of planned behaviour (Ajzen 1991), behaviour is the result of a conscious intention, shaped by people's attitudes, social norms and perceived behavioural control. However, the link between intention and actual behaviour appears to become much weaker in situations where habits take over. Habits can be defined as "a form of automaticity in responding that develops as people repeat actions in stable circumstances" (Verplanken et al. 2006, 91). The "habit discontinuity" hypothesis (Verplanken et al. 2008) states that (travel) habits or routines may become weakened – and hence, reconsidered – if important 'contextual discontinuities' or 'life events' occur, either in the individual's own life (e.g. moving home, changing jobs, marriage or divorce, acquiring a driving licence) and/or in the wider societal (social, economic, spatial) context. Such events may be planned or unplanned, permanent or temporary. Verplanken et al. also suggest that context discontinuity may be coupled with 'self-activation' of personal views or attitudes, e.g. environmental consciousness, resulting in more sustainable travel behaviour.

In order to fully understand the importance of life events and their effect on people's (travel) behaviour, a long-term perspective can be useful. Until the end of the 20th century, travel behaviour research was mostly limited to the use of cross-sectional (instead of longitudinal) data and – with a few exceptions – little attention was paid to the effect of long-term decisions on travel behaviour (Lanzendorf 2003). To address these limitations, 'biographical research' has emerged as an interesting and promising approach to better understand travel behaviour and, more specifically, travel behaviour changes over the life course of individuals. The main advantage of such an approach is that it goes beyond the analysis of the current situation (i.e. which factors determine travel behaviour choices) and also takes into account the temporal dimension (i.e. how people's travel behaviour changes over time and how this can be linked to past experiences as well as future aspirations and plans).

The life course perspective in fact originated as early as in the 1960s, as an interdisciplinary and holistic approach to study the life histories or 'trajectories' of individuals and groups over time and the effects of personal but also social-historical context and conditions on behaviour. It has since then become a flourishing field of research. Within life course theory, a number of useful concepts have been developed that can also be applied to travel behaviour studies. Elder et al. define social pathways as structured trajectories of education and work, family and residences followed by individuals and groups that are shaped by historical forces and social institutions. Within such trajectories (i.e. sequences of roles and experiences), transitions (i.e. changes in state or role) open up a window of opportunity for behavioural change and may lead to a turning point – a substantial change in the direction of one's life, whether subjective or objective (Elder et al. 2003).

Drawing upon the life course approach, Lanzendorf (2003) introduced the term 'mobility biography' to refer to "the total of the longitudinal trajectories in the mobility domain". Lanzendorf considers that even though travel behaviour is to a large extent habitual – habit is understood as "the repeated performance of behaviour sequences by individuals" (Gärling and Axhausen 2003) – it may change over time. Sometimes, this is immediately and sometimes time lagged – as a result of specific events that involve major changes in the life course in either the 'lifestyle' domain (demographic, professional and leisure 'careers'), the 'accessibility' domain (locations of residence, workplace, shopping, leisure and other activities) and/or the 'mobility' domain (availability of modes: car ownership, public transport season ticket and actual activity and travel patterns) (Lanzendorf 2003).

In the conceptual model developed by Chatterjee et al. (2013), turning points in travel behaviour are triggered by contextual change (a life transition event or a change to the external environment). The reasoning is that life transition events can alter the roles people perform, their values and preferences, the resources available for travel and the context for travel (activity space). Intrinsic motivations, facilitating conditions in the external environment and personal history (past – positive or negative – experiences) also play a role as mediating factors (Chatterjee et al. 2013).

Empirical studies confirming the impact of life events on travel behaviour

Since 2003, a growing number of empirical studies have explored and confirmed the impact of life events on travel behaviour. A review of existing studies (Chatterjee and Scheiner 2015; Clark et al. 2014) shows that, indeed, the occurrence of important life events increases the likelihood of (lasting) changes to travel behaviour. Clark et al. also indicate that certain life events tend to cluster together, particularly in early adulthood, which increases their impact on travel behaviour (Clark et al. 2014). It should be noted that most studies conducted so far have been limited in scope, mainly because they have had to rely on relatively small sample sizes that may not be representative of the general population. In recent years, the increased availability of large-scale panel data sets has enabled researchers to track and uncover how travel behaviours evolve over time, for a large sample of test persons.

Two large-scale research projects are discussed in more detail below: the USEmobility survey (2011) and the Life Transitions and Travel Behaviour project in the UK (2012–2014).

The USEmobility survey

In 2011, the "USEmobility" (Understanding Social behavior for Eco-friendly multimodal mobility) project investigated individual reasons that lie behind selecting a mode of transport. More specifically, USEmobility surveyed over 10,000 'swing users', i.e. citizens who had modified their mobility mix in the last five years, from six European countries (Austria, Belgium, Croatia, Germany, Hungary and Netherlands), about the reasons for their modal choice. USEmobility came up with a range of interesting new insights that are summarised below.[1]

The USEmobility survey first showed a lot of **dynamism in people's choice of transport mode**. Almost half of the people addressed in the survey were identified as so-called 'swing users', i.e. reported a change in their use of transport modes in the last five years. In metropolitan areas, a general increase in public transport use was registered, whereas in the rural areas, the change rather tended towards an increased use of private motorised transport.

20% of the people involved in the research had decided to increase their use of public transport or to start using it for the first time. Within this group, the biggest segment (almost 1/3) consisted of 'complete changers' from motorised individual transport to public transport.

The highest dynamic was found for the travel purpose 'way to work'. In 2/3 of the cases, the 'swing users' changed their travel behaviour 'step by step'; in 1/3 of the cases, these changes took place 'overnight'. Swing users' behavioural patterns were also found to be much **more multimodal and much more pragmatic** than initially expected.

On average, 70% of the 'swing users' already used multimodal means of transport. It is also interesting to see that (in all countries) on average more than a quarter of swing users – the most important group – took a pragmatic point of view when choosing their mode of transport, i.e. they made different decisions according to the situation they were in and were the most dynamic in their behavioural patterns (Figure 3.1).

In general, pull-in factors (attractiveness of the transport offer) have a higher relevance than push-out factors (dissatisfaction with the means of transport used so far). We should also note how changes in one's personal situation are more relevant for a change towards public transport than for a change away from public transport. However, a decrease in public transport use is relatively often influenced by dissatisfaction (push-out).

For a continued use (including among swing users who have access to a car), public transport needs to be attractive (pull-in) in comparison to other means of transport. A new and surprising insight of the USEmobility project was how strongly **changes in people's personal situation influence changes in their choice of means of transport.** Indeed, habits and mobility routines play an important role in people's daily mobility.

Changes in one's personal situation (relocation, a new job, birth of children) give people an impulse to rethink their mobility routines, consider

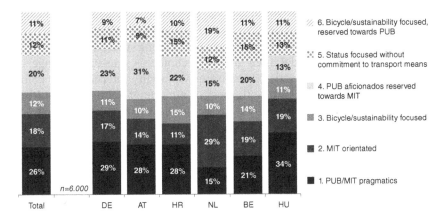

Figure 3.1 USEmobility segments of attitude by country.
Source: Knuth 2012.

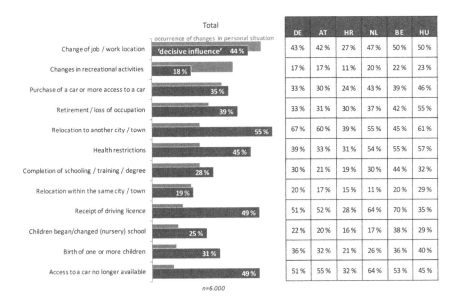

Figure 3.2 Degree of influence of changes in one's personal/private situation on mobility decisions.
Source: Knuth 2012.

alternatives and change their behaviour. Over half of the survey participants stated that change in their personal situation was a central motivation for their reorientation (Figure 3.2).

A change of job or work location had, overall, the highest impact, followed (with considerable decrease in relevance) by relocation to another city or town (very relevant but not that common), increased availability of a car, retirement or loss of occupation and health restrictions (very relevant but not that common). Lost access to a car and obtaining a driving licence are highly relevant factors as well, but rather rare among swing users.

When people reconsider their mobility choices – due to a change in their personal situation and/or because new mobility alternatives become available – **pull-in factors** (attractiveness of a mode of transport, resulting in more frequent use) **and push-out factors** (dissatisfaction with a mode of transport, resulting in less frequent use) move into the spotlight. An overview of influencing factors is presented in Figure 3.3.

The USEmobility survey showed that 'hard' factors (reachability, cost, journey time, waiting time, number of transfers, frequency of connections) had the highest relevance in both the decision to use public transport and multimodal transport more often and, on the contrary, to quit public transport. For users to continue using public transport, their expectations regarding these 'hard' factors need to be fulfilled. If not, it can be expected that users will reduce or cease their use of public transport. Direct connections (without transfers) push increased use of public transport considerably.

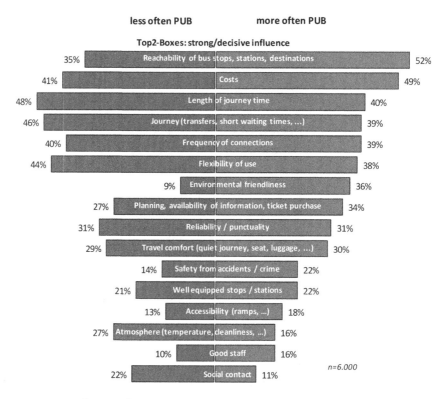

Figure 3.3 Influence of primary push-factors concerning public transport.
Source: Knuth 2012.

On the other hand, inadequate reachability of stops, stations and destinations, long waiting times, insufficient frequency of connections and crowding are the main factors that cause swing users to stop using public transport or use it less often. The influence of so-called 'soft' factors was less pronounced but still considerable. Among the soft factors, flexibility, planning effort, availability of information and environmental friendliness had the highest relevance followed by comfort of travel, atmosphere on the journey and staff. This to say that 'soft' factors can be regarded as complementary success factors that can contribute under certain circumstances to the increased use of public transport; however, they cannot completely substitute the 'hard' factors.

What we should keep in mind is how people facing a change in their personal situation are more open to reconsidering their mobility needs and solutions. Therefore, **directly addressing people whose life circumstances have changed** – for example, providing welcome or info packages to people who have relocated to another city or town or finished school or university; changed their job or retired; families who recently had a baby – opens up significant potential for influencing their modal choice (towards public transport and multimodality) at a relatively small cost.

And even if there is no change in someone's personal situation, the **availability of information and knowledge about the existing (public transport, multimodal) offer** is quite important, as most swing users base their decision to use public transport on their own experience or on information they have received from personal contacts (family, friends, acquaintances, colleagues). Relevant and comprehensive information should be provided to (potential) users, taking into account the needs of different target groups (e.g. young, elderly, disabled users, different travel purposes), e.g. by means of mobility trainings, awareness-raising campaigns.

The life transitions and travel behaviour project

The Life Transitions and Travel Behaviour project (2012–2014) was an 18-month research project conducted by the University of the West of England, the University of Essex and the Department for Transport. The research team made use of 'Understanding society' and British Household Panel Survey (BHPS) data to study how life transitions affect people's travel behaviour over time. The main results are summarised below.

A first important observation is that many people experience life-changing events. In 2009–2010, the most common life events experienced in England were residential relocation (6.9%), change of employer (6.2%), entering employment from non-employment (5.1%), lost employment (3.3%), birth of a child (3.1%) and gaining a driving licence (2.5%) (Life Transitions and Behaviour Study 2014a).

The number of cars in a household is more prone to change at the time of life events: notably, starting and ending cohabitation, getting a driving licence, moving in and out of employment, birth of a child, changes in income, changing employer and residential relocation. Of course, apart from experiencing life events, other (more static) circumstances also play a role, such as household size, the presence of children or access to employment by public transport (Life Transitions and Behaviour Study 2014a).

Similarly, changes in commuting mode were found to be much more likely for those experiencing life events, notably changing job and moving home, because these have an obvious impact on the travel distance and transport options available. No less than 20% of the panel members involved in the study changed commuting mode. It should be noted here that car commuting appears to be far more stable than commuting by any other type of transport: on average, people commute six years by car and only three years by public transport, cycling or walking; and switches to car commuting are far more prevalent than switches to non-car and active commuting. Residential context (density, availability of public transport) also plays a role. Getting a driving licence makes a switch to car commuting more likely, whereas willingness to protect the environment increases the probability of switching from car to other commute modes. Interestingly, a change to car commuting is much more likely (30 times more likely) if the commuting distance increases above two miles, whereas a change to non-car commuting is more likely (but only nine times more likely) if the distance reduces below three miles. Also, environmental attitudes play a role: car commuters who are 'willing to act to protect the environment' are more likely to switch to non-car commuting (Life Transitions and Behaviour Study 2014b).

Conclusion

As most travel behaviour is habitual, it is a challenge to encourage people to consider other transport modes such as public transport, walking or cycling. For policy makers and transport service providers, it follows that life events or transition points can – and should – be considered as opportunities for triggering behavioural change, i.e. for promoting alternative transport options and nudging people towards using them. This could be achieved by means of interventions such as information and awareness raising campaigns, offering free public transport tickets targeted at people who have experienced a life-changing event. Young adults experience more change (move home, start a job, start a family, acquire a driving licence) and are hence an important group to target.

Pilot projects in this direction have already shown promising results. In Copenhagen, when a large sample of commuters who owned a car received a free public transport travel card, public transport use increased – but only among those who had moved home or changed workplace in the last three months (Thøgersen 2012). In the framework of the SEGMENT project (2013), welcome packs and cycle maps were sent to new residents of Utrecht – which led to a modal shift of 4% from car to cycling and public transport. In the West of England, travel advisors from TravelWest visited new residents at home to give them a Travel Information Pack and access to a range of free offers and promotions such as free bus tickets, cycle training and route planning assistance (Travelwest n.d.). Even though at this moment, more research is definitely needed on the efficacy of such interventions and how to measure their success (Chatterjee et al. 2015), we can conclude that taking life events into account should be a key element in any transport policy that aims to achieve travel behaviour change.

Note

1 For the full survey results, see USEmobility D3.6 – Factors influencing behav- ioural change towards eco-friendly multimodal mobility (Knuth 2012).

References

Adjei, Eric, and Roger Behrens. 2012. "Travel behaviour change theories and exper- iments: A review and synthesis." In *31st Southern African Transport Conference*, Pretoria, 9–12 July 2012.

Ajzen, Icek. 1991. "The theory of planned behavior." In *Organizational Be- havior and Human Decision Processes*, 50, no. 2: 179–211. https://doi. org/10.1016/0749-5978(91)90020-T.

Chatterjee, Kiron, and Ben Clark. 2015. "The facts are clear: Life events change travel behaviour. Policy-makers please take note." In *Local Transport Today*, 679: 18. http://eprints.uwe.ac.uk/26972. Accessed 20 January 2020.

Chatterjee, Kiron, and Joachim Scheiner. 2015. "Understanding changing travel be- haviour over the life course: Contributions from biographical research." In *14th International Conference on Travel Behaviour Research*, Windsor, UK, 19–23 July 2015. http://eprints.uwe.ac.uk/28177, Accessed 20 January 2020.

Chatterjee, Kiron, Henrietta Sherwin, Juliet Jain, Jo Christensen, and Steven Marsh. 2013. "A conceptual model to explain turning points in travel behaviour: Application to bicycle use." In *Transportation Research Record*, 2322 (1): 82–90. https://doi.org/10.3141/2322-09.

Clark, Ben, Kiron Chatterjee, Steve Melia, Gundi Knies, and Heather Laurie. 2014. "Life events and travel behaviour: Exploring the interrelationship using UK household longitudinal study data." In *Transportation Research Record*, 2413 (1): 54–64. https://doi.org/10.3141/2413-06.

Elder, Glen H., Monica Kirkpatrick Johnson, and Robert Crosnoe. 2003. "The emergence and development of life course theory." In *Handbook of the Life Course*, edited by Jeylan T. Mortimer and Michael J. Shanahan, 3–19. London: Springer.

Gärling, Tommy, and Kay W. Axhausen. 2003. "Introduction: Habitual travel choice." In *Transportation*, 30: 1–11. https://doi.org/10.1023/A:1021230223001.

Knuth, Klaus-R. 2012. "Deliverable D3.6. Factors influencing behavioural change towards eco-friendly multimodal mobility." USEmobility Project. https://www. levego.hu/en/campaigns/usemobility/reports/survey/. Accessed 20 January 2020.

Lanzendorf, Martin. 2003. "Mobility biographies. A new perspective for under- standing travel behaviour." In Anon. 10th International Conference on Travel

Behaviour Research, Lucerne, 10th August 2003, 1–20. The International Association for Travel Behaviour Research.

Life Transitions and Travel Behaviour Study. 2014a. "Evidence summary 1 – Household car ownership and life events." https://travelbehaviour.com/outputs-lttb/. Accessed 20/01/2020.

Life Transitions and Travel Behaviour Study. 2014b. "Evidence summary 2 – Drivers of change to commuting mode." https://travelbehaviour.com/outputs-lttb/. Accessed 20 January 2020.

"SEGmented Marketing for ENergy efficient Transport (SEGMENT)." n.d. https://ec.europa.eu/energy/intelligent/projects/en/projects/segment. Accessed 20 January 2020.

Thøgersen, John. 2012. "The importance of timing for breaking commuters' car driving habits." In *Collegium*, 12: 130–140. https://helda.helsinki.fi/bitstream/handle/10138/34227/12_08_thogersen.pdf?sequence=1. Accessed 20 January 2020.

Travelwest. n.d. "Moving home. New home, new start, new ways to travel…" https://travelwest.info/movhome. Accessed 20 January 2020.

Van der Waerden, Peter, Harry Timmermans, and Aloys Borgers. 2003. "The influence of key events and critical incidents on transport mode choice switching behaviour: A descriptive analysis." *Paper Presented at the 10th International Conference on Travel Behaviour Research*, Lucerne, August 2003.

Verplanken, Bas, and Wendy Wood. 2006. "Interventions to break and create consumer habits." In *Journal of Public Policy and Marketing*, 25 (1): 90–103. https://doi.org/10.1509/jppm.25.1.90.

Verplanken, Bas, Ian Walker, Adrian Davis, and Michaela Jurasek. 2008. "Context change and travel mode choice: Combining the habit discontinuity and self-activation hypotheses." In *Journal of Environmental Psychology* 28, no. 2: 121–127. https://doi.org/10.1016/j.jenvp.2007.10.005.

Part II

Geographies of mobility poverty

4 The spatial dimension of mobility poverty

Tobias Kuttler

Abstract

Mobility poverty is intrinsically linked to spatial dynamics. Geographies and spatial systems influence how people move. Conversely, movements of all kinds have an impact on locations. They shape the built environment by adding to it the necessary infrastructure for mobility. Mobilities also co-create spatial typologies with particular mobility-related characteristics, such as car-dependent suburban neighbourhoods. Similarly, experiences of mobility disadvantage may be shaped by spatial disadvantages, e.g. by peripheral location in the spatial system. However, the relationship between mobility poverty, spatial factors and social exclusion is never a determinism or automatism, as will be shown. Therefore, this chapter will first provide an overview of spatial systems and dynamics in Europe before turning towards specific challenges in the urban and rural context.

Mobility and the spatial system of Europe

As outlined in the introduction to this volume, the study of unequal mobilities is closely linked to understanding uneven spatial development. Therefore, it is important to understand how these mobilities and geographies interact, with the aim of identifying and analysing the spatial specificities and characteristics of mobility poverty, taking into consideration the geographic elements of dwelling and moving.

The spatial conditions of mobility are determined by several factors. Density, location in the rural-urban network and accessibility are the spatial factors that have a strong impact on individual mobility behaviour and needs as well as the ability to move (motility). Hence, an analysis of the spatial conditions of mobility poverty needs to consider spatial dynamics in Europe. Urban expansion, suburbanisation and reconfiguration of the European spatial system have resulted in an urban-rural transition zone, referred to here as the peri-urban areas. These areas are very dynamic and are themselves in a process of constant transformation, posing taxonomic problems in their definition. They exhibit demographic, socio-economic and other characteristics that are different from purely rural and urban areas, with specific impacts on individual mobilities and mobility systems. Indeed, suburbanisation and peri-urbanisation (as well as the parallel processes of re-urbanisation) pose a substantial challenge to defining what is urban and what is rural today. Cities in Europe and all over the world have grown beyond their former city limits, a circumstance that makes it necessary to fully understand the implications of definitions such as "rural", "urban" and "peri-urban".

Keeping in mind the challenge of definitions among geographical arrangements, it is true that geography and mobility are linked to each other in myriad and complex ways. The spatial distribution of human activities, movement of people and goods between places and spatial characteristics interact with and inform each other. They form a system that is under constant transformation. Far from being direct causal relationships and determinants, the relationship between geography and mobility is often subtle and therefore often not clearly understood. Too often, development projects – seemingly well executed and based on the principles of integrated planning – have failed, challenging some of the fundamental assumptions about the interaction between space and mobility.

Density is one of the concepts that are often employed when trying to understand the relationship between space and mobility (Frey and Zimmer 2000). In the past, the density of human activities in a certain location certainly shaped evolution and the basic conditions of how people travel on an everyday basis. Conversely, the conditions of transport systems have influenced locational choices. A high density of people, activities and opportunities justifies sophisticated transportation infrastructure and networks as well as a high frequency and speed of transport services. High capacity transport systems are a prerequisite for maintaining and fostering the competitiveness of an economic location. Accessibility is thus most advanced in regions with a high density of population and economic activity.

When living and working was still located in the same place or close to each other, the need for more elaborated modes of transport was not universal. The limits of walking distance shaped early towns and settlements in Europe with their characteristic densities and associated benefits, but also had negative ramifications. With sustainability having become the primary guiding principle for spatial development, reference to Europe's spatial history is often made when envisioning the ideal city in terms of size and density. With the renaissance of non-motorised mobility in cities across Europe, dense and compact urban development is once again one of the primary objectives in spatial planning. A compact urban structure combined with a spatial system that is characterised by a hierarchical structure of equally developed main urban hubs with regional centres and smaller settlements is conducive to the use of public and non-motorised transportation.

This chapter will start with the most important spatial dynamics in the more recent history of Europe, i.e. suburbanisation and peri-urbanisation. As mentioned above, changing locations of living, work, leisure, etc. and the way they are separated from each other – or concentrated in one place – substantially influence everyday mobility. It will be shown that local, regional as well as global factors have an impact on the intertwined processes of urban expansion and concentration. After having outlined these spatial dynamics in Europe, a better understanding of spatial definitions and categories will be provided. Next, we employ a European macro perspective to illustrate the issue of accessibility to vital urban functions before closing this chapter with a short discussion on commuting.

Suburbanisation and peri-urbanisation in Europe

Residential suburbanisation in Europe has resulted in a separation of workplaces and places of living with larger distances being travelled on an everyday basis. However, the fact that the historic core cities remained major centres of employment, supply and culture means that a substantial part of personal travel is still directed towards the urban centres, supporting public transport use along main corridors. Unlike Europe, other parts of the world have seen a much more dispersed spatial pattern of suburbanisation that was – and still is – overwhelmingly car-based. Furthermore, in many places, residential suburbanisation was followed by suburbanisation of work, leisure and shopping (Hall 2003; Hart 2000).

The characteristic result of suburbanisation in Europe is a circle of wealthy rural communities around a core city, characterised by urban professionals living there and commuting to the urban core (Nelson and Sanchez 1999, 689). Apart from urban professionals, industries and services have been relocating to rural areas, especially in the second half of the 19th century, when there was a strong tendency towards urban deconcentration and counter-urbanisation (Ravetz, Fertner and Nielsen 2013, 17). Ex-urban retirement settlements are another form of suburb that can be found for example in Spain (Zasada et al. 2010), also second and holiday homes.

However, suburbanisation is only one of the processes that caused urban expansion and differentiation of Europe's historic spatial system: there has been an outward process from the city to the fringe, what is usually understood as suburbanisation. While the relocated households, commerce, industries, services and entertainment usually remain in close functional connection with the core city, more recently a variety of spatial uses has been established outside core cities. These types of uses may be in spatial proximity to a certain city, but are functionally more closely linked to global networks, such as industrial production chains. Such diverse and often contradictory spatial processes are subsumed under the term "peri-urbanisation" in contemporary literature. This term describes spatial development taking place in "transition zones" that are considered "something in between", neither fully urban nor rural (Allen 2010; Ravetz, Fertner and Nielsen 2013).

Peri-urban regions are the most dynamic regions of urbanised Europe and they exhibit a range of – sometimes conflicting – transport-related characteristics and mobility requirements. Still, there is much confusion about the geographical extent of the peri-urban: often, the peri-urban is known as the peri-urban fringe (e.g. Errington 1994). Peri-urbanisation is used in combination with other terms like urban sprawl, suburbanisation, ex-urbanisation and re-urbanisation (Gant, Robinson, and Fazal 2011); also, in different academic traditions, different terms are used to describe similar processes, for example in Germany and France (Forum Mobile Lives 2013).

That being said, authors generally agree anyway that the peri-urban, or the urban fringe, is the zone between the urban and the rural that is under transformation (e.g. Council of Europe 2007; Douglas 2012; Zasada et al. 2011). The term peri-urban thus describes less a specific territory, but rather a process of incremental and "incomplete" urbanisation. Ravetz, Fertner and Nielsen (2013, 13) have highlighted that the peri-urban can be seen not just as a zone of transition, but rather "a new kind of multi-functional territory".

However, when peri-urbanisation is understood as a process, the challenge remains how to define the peri-urban geographically. Ravetz, Fertner and Nielsen (2013) have suggested the term "rural-urban region" and an associated comprehensive methodology to define the peri-urban geographically, based on an extensive literature review. They identify two spatial types (Ravetz, Fertner and Nielsen 2013, 18–19):

1   Urban fringe: a zone along the edges of the built-up areas, which comprises a scattered pattern of lower density settlement areas, urban concentrations around transport hubs, together with large green open spaces, such as urban woodlands, farmland, golf courses and nature reserves;
2   Urban periphery: a zone surrounding the main built-up areas, with a lower population density, but belonging to the Functional Urban Area; this may include smaller settlements, industrial areas and other urban land uses within a matrix of functional agriculture.

The suburban areas in this classification are part of the urban, built-up area. Suburban areas are generally lower density contiguous built-up areas, which are attached to inner urban areas, and where houses are typically not more than 200 m apart, with local shops and services, parks and gardens (Figure 4.1).

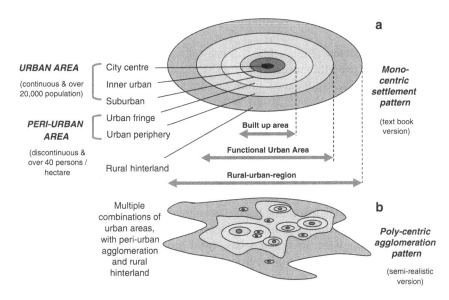

Figure 4.1 The peri-urban area as a part of the rural-urban region.

Source: Ravetz, Fertner and Nielsen 2013, 18. Reprinted by permission from Springer Nature: The Dynamics of Peri-Urbanization by Joe Ravetz, Christian Fertner, Thomas Sick © 2013.

Changing urban systems

The brief review of European peri-urban development suggests that dwelling dynamics are proceeding in parallel, mutually enforcing each other, but leading to contradictory outcomes. Spatial relations increasingly do not follow a conventional centre-periphery model (as, by the way, they never fully have in the past). In the contemporary globalised world, locations are linked to each other in ways that are territorially contiguous, by the everyday movement of people and goods; at the same time, places are linked to others elsewhere in spatially non-contiguous ways through virtual mobility and long-distance travel. As Manuel Castells argued, the logic of the "space of flows" is dominant in contemporary life and is associated with political and economic power (Castells 1999).

Additionally, the dominance of some urban centres in Europe has led to substantial negative agglomeration effects that fostered new town development across Europe. Such developments have resulted in metropolitan regions being more polycentric with new towns becoming hubs of employment themselves. For transport, this has resulted in the need for networks that allow tangential connectivity avoiding the central city.

Ravetz, Fertner and Nielsen (2013, 21–26) analysed different local, regional and global factors and causes that have impacts on Europe's spatial system. They identified three different types of urban growth and dynamics that cause the city to grow beyond its boundaries:

- **Urban expansion** (Figure 4.2): this is foremost a result of population and economic growth, causing a higher demand for housing and commercial areas. Transport accessibility to employment and services, as well as the attractiveness of the environment and land values then determine the new locations for housing. Physical and political constraints also play a role. Furthermore, housing demand is affected by a decrease in average household size, but also a higher demand for residential space in general. Economic and employment growth and changing employment patterns further drive urban expansion by an increase in the building stock and land-use conversion. More in the United States than in Europe, urban expansion has been driven by dependency on the automobile, creating built-up landscapes centred around cars.
- **Regional agglomeration and urban–rural linkages** (Figure 4.3): dynamics on an inter-urban and regional scale are constantly reshaping spatial relationships, resulting in inter-urban or regional agglomerations. Single cities are replaced by regional urban systems of inter-connected and polycentric types of settlement. Processes that occur in rural areas, such as economic restructuring, land market changes and agricultural modernisation, can also support agglomeration dynamics.

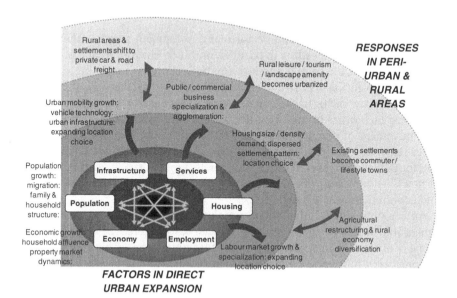

Figure 4.2 Processes of urban expansion.
Source: Ravetz, Fertner and Nielsen 2013, 22. Reprinted by permission from Springer Nature: The Dynamics of Peri-Urbanization by Joe Ravetz, Christian Fertner, Thomas Sick © 2013.

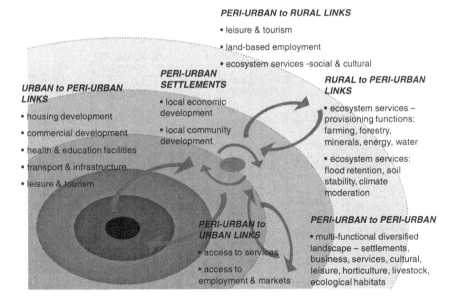

Figure 4.3 Agglomeration development and rural-urban linkages.
Source: Ravetz, Fertner and Nielsen 2013, 24. Reprinted by permission from Springer Nature: The Dynamics of Peri-Urbanization by Joe Ravetz, Christian Fertner, Thomas Sick © 2013.

- **Global-local restructuration** (Figure 4.4): the main dynamic driving this process is globalisation. Globalisation has multiple effects, such as economic effects on business structures and finance; it also has political effects on the urban systems and hierarchies of nation states and Europe as a whole and, finally, cultural effects through the media and information and communications technology (ICT) can also be observed. All these processes shape urban dynamics – growth as well as decline – in combination with the aforementioned forces. Associated processes, such as privatisation and franchising, have far-reaching impacts on governance and public services. New forms of consumption, leisure and tourism have reshaped spaces far beyond traditional urban areas. In contrast to globalisation, there are also localisation processes, resulting in new cultural identities, new forms of enterprises and diverse use of spaces.

Spatial categories and their challenges

As outlined above, the challenges that individuals encounter while being mobile are directly linked to the characteristics of the space in which they move. Furthermore, social disadvantages affect people differently in rural, peri-urban and urban areas. While the rural-urban differentiation seems

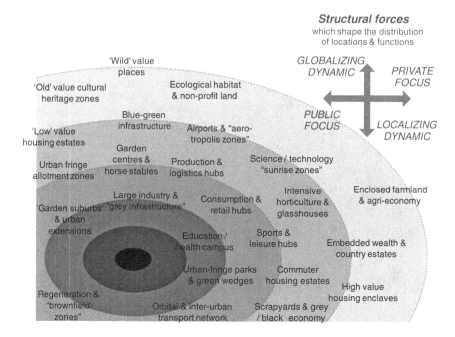

Figure 4.4 Global-local restructuration.
Source: Ravetz, Fertner and Nielsen 2013, 26. Reprinted by permission from Springer Nature: The Dynamics of Peri-Urbanization by Joe Ravetz, Christian Fertner, Thomas Sick © 2013.

obvious at first sight, as illustrated above, it is often difficult to say where the city ends and a rural area begins. In many regards, today peri-urban and rural areas exhibit the same characteristics as urban areas. Therefore, rather than separating spaces, it is conceptually helpful to understand how different urban and rural spaces are always linked to each other by movements of any kind.

Economic characteristics of a given area are often used to define the urban and the rural. From this perspective, the urban is usually identified by its economic activities being mostly non-agricultural. Moreover, the diversity of economic activities in services and production is considered in urban areas, while economic activities in rural areas are usually less heterogeneous. Another way of separating the urban and the rural is to identify areas that are related to urban cores in a functional way, which are often economic links. A functional definition that is related to a city's economy is the territory delimited by the commuting interrelations around a core city (Frey and Zimmer 2000).

Definitions of metropolitan areas take a similar approach by identifying a contiguous area that is under the primary influence of an urban core, which encompasses more and less dense areas. Lastly, urban areas can be differentiated from rural ones by their degree of "urbanism", meaning the social characteristics of a settlement. In this vein, it is usually argued that urban lifestyles differ from rural ones and so do people's worldviews, values and behaviours (Frey and Zimmer 2000).

Such categorisations highlight that the urban and the rural exist as a continuum. For example, considering the territorial limits of functional links to and from a city, it must indeed be recognised that for a long time agricultural production in rural areas has served to sustain the increasing population living in cities. Seen from that functional perspective, most agricultural space is an extended urban space. Furthermore, in industrialised countries, urban lifestyles have become prevalent in densely as well as sparsely populated areas, not least due to the opportunities of ICT. On the other hand, in the Global South, many rural migrants living in cities maintain traditional rural lifestyles. Consequently, several academics have argued that there is no longer anything beyond the urban, declaring that the contemporary world is characterised by "planetary urbanism" (Brenner and Schmid 2015).

European countries have historically used different methodologies to classify their spaces and settlements. This has the effect that a settlement which is called a city in one country would be regarded as a village in another. For example, in Denmark, a settlement is called a city when the population is above 200 inhabitants in a contiguous built-up area with distances between houses of not more than 200 metres (Statistics Denmark 2014). In Germany, the urban category is defined by population size and its position in the hierarchy of supply centres. The smallest urban settlement starts with a population of 5,000, given the fact that this settlement is a basic supply centre to the region. However, where a settlement has fewer than 5,000 inhabitants but is categorised as a basic supply centre, that settlement is a "rural town" (Bundesinstitut für Bau-, Stadt- und Raumforschung 2015). These examples show that, although both countries include population size as a measure to classify settlements, they can hardly be compared based on these definitions.

Urban functions and accessibility in Europe

Accessibility is of crucial importance when analysing mobility poverty. People on low income and those experiencing social disadvantage often experience below-average accessibility. Poor accessibility leads to fewer opportunities of social interaction and fulfilment (e.g. health or educational services), with direct impact on a person's quality of life and well-being. As a result, people living in such areas often need to resort to private car use to fulfil their mobility needs. Páez et al. (2009) studied the transport accessibility limitations of three vulnerable segments – the elderly, those on low income and single parents – regarding three activities – accessing health care, food services and jobs – respectively. The study recorded low accessibility for the three groups, even in the case of those persons owning (or using) a car. The study makes evident that the regions where vulnerable segments live exhibit a lower density of relevant activities (compared with other areas), requiring the people living in such a region to travel for longer or preventing them from accessing such activities.

Such findings highlight the necessity to shed light on accessibility in different European regions, especially those regions that exhibit low population density and low density of opportunities that need to be accessed in everyday life. Opportunities for basic everyday needs are mostly concentrated in regional towns and cities, while locations that cater for specialised needs can be found in larger cities. There are substantial differences across Europe in terms of time needed to reach regional towns and cities, posing a disadvantage to those living in low-density, remote areas.

Figure 4.5 shows the accessibility of cities larger than 50,000 inhabitants within 60 minutes of road travel time while Figure 4.6 shows the same for 60 minutes of rail travel time.[1] From most locations in Western and Central Europe, at least one regional city can be reached by road within 60 minutes; from many places, even more than ten regional cities. Accessibility by road and rail is thus highest in the centre of Europe. Highly urbanised parts of the United Kingdom, the Netherlands, Belgium and Germany (e.g. the Ruhr region) have the best accessibility by road and rail in Europe. Also, in and around cities in western and eastern France, many parts of Germany, the north of Italy and some parts of Spain, accessibility by road and rail is high.

The analysis of accessibility by road and rail highlights those regions in Europe that do not have access to urban functions in reasonable time. These are regions like Mecklenburg-Western Pomerania (Germany), many parts of France and Spain and areas in Poland and the Czech Republic. For rail, the extent of these areas is even bigger in almost all countries (Spiekermann et al. 2013, 113–116).

To understand how regional accessibility and mobility poverty are related to dynamics at macro level, the impact of globalisation on European regions needs to be addressed. This is also necessary because the concentration of economic activity, employment opportunities and other vital functions in larger cities is likely to continue in the future, while at the same time the disconnection of peripheral regions is continuing, reducing economic activity, employment opportunities and quality of life in these regions (Martinez-Fernandez et al. 2012).

Globalisation and the development of cities with Global City status have greatly impacted accessibility levels in Europe. Globally connected cities like London, Paris, Amsterdam and Frankfurt benefit from their well-developed urban transport systems, their integration into high-speed rail networks as well as their international airports allowing direct connections to other world cities. While these cities benefit from their above-average accessibility levels, smaller cities struggle to reach similar levels of global and inter-metropolitan connectivity (Spiekermann et al. 2013, 77–80).

**Availability of urban functions (2011):
Number of cities > 50,000 inhabitants within
60 minutes road travel time (raster level)**

- 0
- 1
- 2
- 3 - 5
- 6 - 10
- 11 - 25
- 26 - 50
- 51 - 76
- no data

Figure 4.5 Accessibility of urban functions (2011): Number of cities >50,000 inhabitants within 60 minutes' road travel time.

Source: Spiekermann et al. 2013, 114.

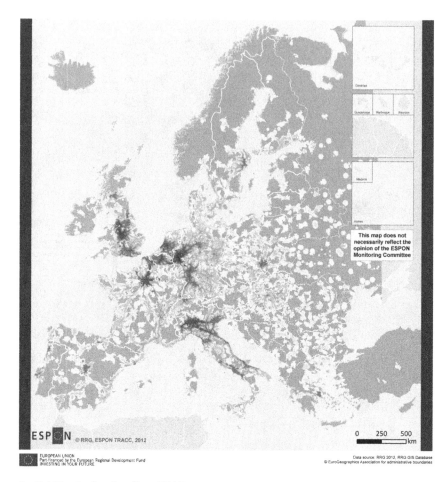

Availability of urban functions (2011):
Number of cities > 50,000 inhabitants within
60 minutes rail travel time (raster level)

0   no data
1
2
3 - 5
6 - 10
11 - 25
26 - 50
51 - 66

Figure 4.6 Accessibility of urban functions (2011): Number of cities >50,000 inhabitants within 60 minutes' rail travel time.

Source: Spiekermann et al. 2013, 115.

As an example of global accessibility, Figure 4.7 highlights travel times to New York from Europe, providing a good impression of how different accessibility levels are in different European regions.

These figures show how Europe's spatial and transport systems have developed around an agglomeration of large cities in central Western Europe

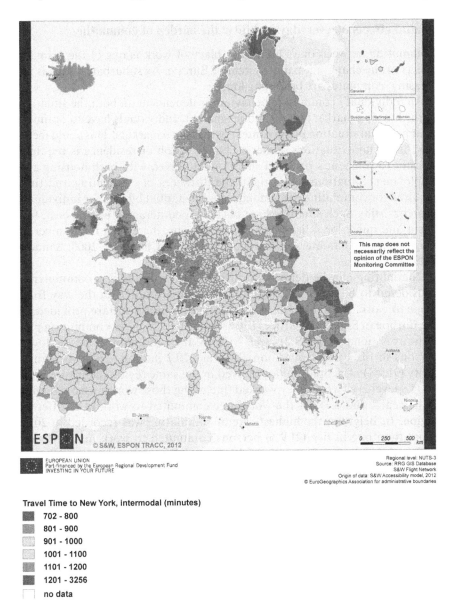

Travel Time to New York, intermodal (minutes)

- ■ 702 - 800
- ▨ 801 - 900
- ▢ 901 - 1000
- □ 1001 - 1100
- ▨ 1101 - 1200
- ■ 1201 - 3256
- □ no data

Figure 4.7 Global accessibility of European Regions with the example of travel times to New York, intermodal (minutes).

Source: Spiekermann et al. 2013, 78.

and, in contrast, how development in peripheral regions of Europe has been at a comparatively lower level. However, Europe's spatial development cannot simply be described as a model of centres and peripheries. Additionally, with the eastern and southern extensions of the European Union, the historical spatial hierarchy is going to be challenged in the future.

Spatial effects on everyday mobility: the burden of commuting

Commuting between one's home and place of work is one of the primary effects of the changing spatial system of Europe. As suburban sprawl is increasing, commutes are becoming longer.

An individual's commuting behaviour is dependent on both the situation of the housing market and the labour market. Individuals have to optimise their personal situation in the context of these two markets. This could mean that, in certain situations, either a change of job or residence is required to cater to someone's personal situation. However, with both housing and employment situations being tight, the distances of commuting and time required for commuting can amount to a substantial burden for individuals (Lorenz 2018). Such situations are partially counteracted by low-cost, frequent and quick long-distance mobility options, increasing the number of long-distance commuters in Europe (Lyons and Chatterjee 2008; Sandow 2019; Viry 2015).

Commuting can be defined in different ways. Usually, a commuter is considered to be a person crossing a municipal border on the way from home to work. Actual numbers on commuting in Europe are provided by the European Statistical Office (Eurostat), which analyses commuting patterns at regional (NUTS 2) level. In 2015, the total number of employed persons in the European Union (28) was 220.7 million. The overall majority (91.9%) of people employed lived in the same region (defined here at NUTS level 2) as where they worked (including those working from home). This means that 8.1% of the workforce commuted to work in a different region. In Belgium, the highest rate of commuting was recorded in 2015. More than one in five (21.9%) persons commuted to work in a different region (NUTS level 2). Also, in the United Kingdom, the Netherlands, Austria and Slovakia, high shares of commuting to a different region can be observed.[2]

Commuting substantially shapes the everyday life of millions of Europeans residents. Not only do European commuters spend a lot of time in cars and trains; commuting is also considered to be a burden and diminishes subjective well-being (Stutzer and Frey 2008; Wurhofer et al. 2015). However, it has also been highlighted that commuting by car in particular is perceived as desirable due to the freedom that a car provides and the potential to enable someone to break free from day-to-day responsibilities (Boyle 2016). Cycling commuters who report dangerous experiences during their regular work commute also highlight their well-being, enjoyment and happiness of doing physical exercise and the ability to wind down after work (Guell et al. 2012).

The analysis of commuting times in Europe illuminates the time-related burden that many Europeans face. With longer commutes, the relationship between commuting and well-being is a prime concern.

A 2015 survey asked for the commuting times of citizens in European member states. The mean commuting time of survey respondents was highest in the UK and Sweden with more than 50 minutes for travelling from home to work and back. The lowest mean commuting times were reported from Italy, Portugal and Cyprus (less than 30 minutes) (Eurostat 2015).

Between 2005 and 2015, the overall mean commuting time in the EU decreased slightly. This is due to contrasting developments in commuting in the member states: while in most of the western and northern member states, mean commuting times increased, in some countries by more than five minutes; commuting times in most eastern and southern member states decreased (see Figure 4.8). In some member states, like Slovakia, Cyprus, Poland and Romania, commuting times decreased by more than 15 minutes (Eurostat 2015).

According to this survey, almost 25% of survey respondents travelled more than one hour from home to work and back (see Figure 4.9). About 2% travelled more than two hours from home to work and back. The highest share of those commuting more than one hour was found in the United Kingdom (35%), including the highest share among member countries of those who travelled more than two hours (5%). After the United Kingdom, similarly high shares of survey respondents reported commuting times

Figure 4.8 Change in mean commuting time 2005–2015 (in minutes).
Source: Authors, based on Eurostat 2015.

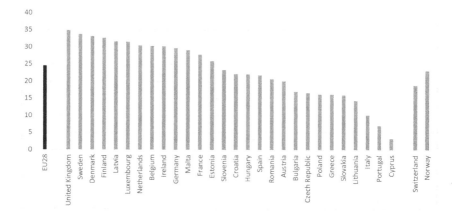

Figure 4.9 Share of employed population with mean commuting time of more than one hour.
Source: Authors, based on Eurostat 2015.

longer than one hour. The lowest shares were observed in Italy, Portugal and Cyprus (10% and lower) (Eurostat 2015).

However, the burden of commuting is not only about the time spent in a mode of transportation per se. Anger and frustration during commuting are a common experience for every commuter. Frustration, anger and stress during a car commute are experienced when unexpected congestion occurs, leading to fear about job-related consequences. Being stuck in traffic congestion is often perceived as a waste of time and a loss of spare time. Also, being stuck in traffic is associated with a loss of control (Guell et al. 2012; Wurhofer et al. 2015). Similarly, longer commuting distances are associated with lower satisfaction with family life and leisure time (Lorenz 2018).

The burden of commuting can become a severe disadvantage when an individual also experiences a disadvantage on the housing and labour market. In the context of mobility poverty, this means that instead of experiencing a low level of mobility, the experience of travelling due to compulsion causes a significant burden. Especially for those with low financial means and limited social and professional networks, any adjustment to optimise the balance between the housing market and labour market is limited. This can lead to substantial distances and time spent travelling that is not compensated by a higher income or a lower rent. Also, people who become employed again after a period of unemployment may start with a less attractive job that includes long journeys to work (Stutzer and Frey 2008).

Furthermore, a study in the United States showed that increased commuting reduces political participation, a phenomenon that can be observed particularly among poor households. Newman, Johnson and Lown (2014, 29) state: "While higher income individuals are not immune from having to commute, [...] the negative effect of commuting with respect to political interest and participation is entirely concentrated among the lower working class".

There are also differences when taking gender into perspective. A study from the United Kingdom highlights that commuting has a higher negative effect on the well-being of women, due to their larger responsibility for day-to-day household tasks, including childcare (Roberts, Hodgson, and Dolan 2011). Although the share of highly educated women in long-distance commuting is larger than male commuters, the majority of women long-distance commuters are low-income earners. This means that men benefit financially more than women from long-distance commuting (Sandow 2019; Sandow and Westin 2010).

Higher commuting burdens are also reported for immigrants and ethnic minorities. The spatial mismatch of housing and job opportunities and its adverse effects (e.g. in the form of a commuting burden) on African American communities in racially segregated cities in the United States has controversially been discussed since the 1960s (Gobillon and Selod 2013). A study from Madrid in Spain showed that immigrants face higher commuting times than the domestic population. This difference is partially a result of residential segregation in certain parts of the city and difficulties in employment accessibility (Blázquez, Llano, and Moral 2010).

Notes

1 Cities with at least 50,000 inhabitants are selected as destinations, assuming that only cities of that minimum size provide a full range of public and private services and functions.
2 These figures are also the effect of the different sizes of the NUTS 2 regions. Especially for Belgium, the Netherlands and the UK, as these countries are densely populated, and the NUTS 2 regions are small in area size. This is confirmed by the observation that in Greece, Spain, Portugal and Romania, commuting across regions is relatively low because the NUTS 2 regions in these countries are very large (Eurostat 2016).

References

Allen, Adriana. 2010. "Neither rural nor urban: service delivery options that work for the peri-urban poor." In *Peri-urban Water and Sanitation Services: Policy, Planning and Method*, edited by Mathew Kurian and Patricia McCarney, 27–61. Dordrecht: Springer, Springer Science & Business Media.

Blázquez, Maite, Carlos Llano, and Julian Moral. 2010. "Commuting times: Is there any penalty for immigrants?" In *Urban Studies* 47 (8): 1663–1686. https://doi.org/10.1177/0042098009356127.

Boyle, Patrick. 2016. "Routine journeys, complex networks: media-centrism, the dispositif of road safety, and practices of commuting by car in everyday Ireland." Doctoral Dissertation, Maynooth University. https://pdfs.semanticscholar.org/a569/b7fc7a2f32e0617736bad14d4e7f5cd61bb1.pdf. Accessed 24 March 2020.

Brenner, Neil, and Christian Schmid. 2015. "Towards a new epistemology of the urban?" In *City* 19 (2–3): 151–182. https://doi.org/10.1080/13604813.2015.101471.

Bundesinstitut für Bau-, Stadt- und Raumforschung. 2015. "BBSR Homepage – Stadt- und Gemeindetyp." https://www.bbsr.bund.de/BBSR/DE/Raumbeobach tung/Raumabgrenzungen/StadtGemeindetyp/StadtGemeindetyp_node.html. Accessed 25 May 2018.

Castells, Manuel. 1999. "Grassrooting the space of flows." In *Urban Geography* 20 (4): 294–302. https://doi.org/10.2747/0272-3638.20.4.294.

Council of Europe. 2007. *Spatial Development Glossary: European Conference of Ministers Responsible for Regional/Spatial Planning (Cemat)*. Bilingual. Territory and Landscape Series v.2. Strasbourg: Council of Europe. https://rm.coe. int/CoERMPublicCommonSearchServices/DisplayDCTMContent?documentId= 09000016804895e5. Accessed 7 June 2018.

Douglas, Ian. 2012. "Peri-urban ecosystems and societies: Transitional zones and contrasting values." In *The peri-urban Interface: Approaches to Sustainable Natural and Human Resource Use*, edited by Duncan McGregor and David Simon, 41–52. Abingdon: Routledge.

Errington, Andrew. 1994. "The peri-urban fringe: Europe's forgotten rural areas." In *Journal of Rural Studies* 10 (4): 367–375. https://doi.org/10.1016/0743-0167 (94)90046-9.

Eurostat. 2015. "How many minutes per day do you usually spend travelling from home to work and back?" https://c-adm.eige.europa.eu/data/view?code=ta_time use_wktime_comm_ewcs_commutetime_mean. Accessed 21 April 2020.

Eurostat. 2016. "Statistics on commuting patterns at regional level." https://ec.europa. eu/eurostat/statistics-explained/pdfscache/50943.pdf. Accessed 24 June 2020.

Forum Mobile Lives. 2013. "Rehabilitating the peri-urban: How to live and move sustainably in these areas?" http://en.forumviesmobiles.org/publication/livres-forum/2013/09/25/rehabilitating-peri-urban-1276. Accessed 10 May 2018.

Frey, William H., and Zachary Zimmer. 2000. "Defining the city." In *Handbook of Urban Studies* edited by Ronan Paddison, 14–35. London: Sage Publications.

Gant, Robert L., Guy M. Robinson, and Shahab Fazal. 2011. "Land-use change in the 'edgelands': Policies and pressures in London's rural–urban fringe." In *Land Use Policy* 28 (1): 266–679. https://doi.org/10.1016/j.landusepol.2010.06.007.

Gobillon, Laurent, and Harris Selod. 2013. "Spatial mismatch, poverty, and vulnerable populations." In *Handbook of Regional Science*, edited by M. M. Fischer and Peter Nijkamp. Berlin, Heidelberg: Springer.

Guell, Cornelia, Jenna Rachel Panter, Natalia R. Jones, and David B. Ogilvie. 2012. "Towards a differentiated understanding of active travel behaviour: using social theory to explore everyday commuting." In *Social Science & Medicine* 75 (1): 233–239. https://doi.org/10.1016/j.socscimed.2012.01.038.

Hall, Peter. 2003. "A European perspective on the spatial links between land use, development and transport." In *Transport and Urban Development*, edited by David Banister, 75–98. Abingdon: Routledge.

Hart, Tom. 2000. "Transport and the city." In *Handbook of Urban Studies* edited by Ronan Paddison, 102–123. London: Sage Publications.

Lorenz, Olga. 2018. "Does commuting matter to subjective well-being?" In *Journal of Transport Geography* 66: 180–199. https://doi.org/10.1016/j.jtrangeo. 2017.11.019.

Lyons, Glenn, and Kiron Chatterjee. 2008. "A human perspective on the daily commute: Costs, benefits and trade-offs." In *Transport Reviews* 28 (2): 181–198. https://doi.org/10.1080/01441640701559484.

Martinez-Fernandez, Cristina, Ivonne Audirac, Sylvie Fol, and Emmanuele Cunningham-Sabot. 2012. "Shrinking cities: Urban challenges of globalization." In *International Journal of Urban and Regional Research* 36 (2): 213–225. https://doi.org/10.1111/j.1468-2427.2011.01092.x.

Nelson, Arthur C., and Thomas W. Sanchez. 1999. "Debunking the exurban myth: A comparison of suburban households." In *Housing Policy Debate* 10 (3): 689–709.

Newman, Benjamin J., Joshua Johnson, and Patrick L. Lown. 2014. "The "daily grind" work, commuting, and their impact on political participation." In *American Politics Research* 42 (1): 141–170. https://doi.org/10.1177/1532673X13498265.

Páez, Antonio, Ruben Mercado, Steven Farber, Catherine Morency, and Polytechnique Montréal. 2009. *Mobility and Social Exclusion in Canadian Communities: An Empirical Investigation of Opportunity Access and Deprivation.* Quebec: Policy Research Directorate, Strategic Policy and Research, Government of Canada.

Ravetz, Joe, Christian Fertner, and Thomas S. Nielsen. 2013. "The dynamics of peri-urbanization." In *Peri-urban Futures: Scenarios and Models for Land Use Change in Europe*, edited by Kjell Nilsson, Stephan Pauleit, Simon Bell, Carmen Aalbers, and Thomas A. S. Nielsen, 13–44. Berlin and Heidelberg: Springer Verlag.

Roberts, Jennifer, Robert Hodgson, and Paul Dolan. 2011. ""It's driving her mad": Gender differences in the effects of commuting on psychological health." In *Journal of Health Economics* 30 (5): 1064–1076. https://doi.org/10.1016/j.jhealeco.2011.07.006.

Sandow, Erika. 2019. "Til work do us part: The social fallacy of long-distance commuting." In *Integrating Gender into Transport Planning: From One to Many Tracks*, edited by Christina Lindkvist Scholten and Tanja Joelsson, 121–144. London: Palgrave Macmillan.

Sandow, Erika, and Kerstin Westin. 2010. "The persevering commuter – Duration of long-distance commuting." In *Transportation Research Part A: Policy and Practice* 44 (6): 433–445. https://doi.org/10.1016/j.tra.2010.03.017.

Spiekermann, Klaus, Michael Wegener, Viktor Květoň, Miroslav Marada, Carsten Schürmann, Oriol Biosca, Andreu Ulied Segui, Harri Antikainen, Ossi Kotavaara, Jarmo Rusanen, Dorota Bielańska, Davide Fiorello, Tomasz Komornicki, Piotr Rosik, and Marcin Stepniak. 2013. "TRACC transport accessibility at regional/local scale and patterns in Europe." https://www.espon.eu/sites/default/files/attachments/TRACC_FR_Volume2_ScientificReport.pdf. Accessed 25 March 2020.

Statistics Denmark. 2014. "Documentation of statistics for urban areas 2014." http://www.dst.dk/Site/Dst/SingleFiles/kvaldeklbilag.aspx?filename=adcefe43-2802-462f-9ba6-38d790b6c6dcUrban_Areas_2014. Accessed 25 May 2018.

Stutzer, Alois, and Bruno S. Frey. 2008. "Stress that doesn't pay: The commuting paradox." In *Scandinavian Journal of Economics* 110 (2): 339–366. https://doi.org/10.1111/j.1467-9442.2008.00542.x.

Wurhofer, Daniela, Alina Krischkowsky, Marianna Obrist, Evangelo Karapanos, Evangelos Niforatos, and Manfred Tscheligi. 2015. "Everyday commuting: prediction, actual experience and recall of anger and frustration in the car." In

Proceedings of the 7th International Conference on Automotive User Interfaces and Interactive Vehicular Applications, 233–240. https://doi.org/10.1145/2799250.27 99251.

Zasada, Ingo, Susana Alves, Felix Claus Müller, Annette Piorr, Regine Berges, and Simon Bell. 2010. "International retirement migration in the Alicante region, Spain: Process, spatial pattern and environmental impacts." In *Journal of Environmental Planning and Management* 53 (1): 125–141. https://doi.org/10. 1080/09640560903399905.

Zasada, Ingo, Christian Fertner, Annette Piorr, and Thomas Alexander Sick Nielsen. 2011. "Peri-urbanisation and multifunctional adaptation of agriculture around Copenhagen." In *Geografisk Tidsskrift-Danish Journal of Geography* 111 (1): 59–72. https://doi.org/10.1080/00167223.2011.10669522.

5 The urban arena

Tobias Kuttler

Abstract

The importance of urban mobility for social inclusion is widely acknowledged. This is why improving mobility in cities has become a
significant policy and development goal worldwide. However, mobility
disadvantage in cities is poorly understood because often simplified assumptions are made about the relationship between spatial dynamics
and social position. This chapter tries to investigate some of the phenomena that have been prevalent in European cities for many decades
and their impact on personal mobility. Starting with a short comparative overview of the state of cities in Europe, this chapter will then turn
towards an examination of different urban processes such as spatial
segregation, urban deprivation, gentrification and centre-periphery
dynamics.

Mobility in cities

Mobility in cities is one of the core concerns of international policy and
development debate. On the one hand, the harmful effect of urban mobility to climate change, air pollution and diminishing quality of urban
life in cities is widely recognised. On the other hand, access to safe and
affordable mobility options is considered a prerequisite for a just, equitable
and inclusive urban development. The urgency of the challenges related to
urban mobility is reflected in the importance attributed to this topic in recent development frameworks such as the New Urban Agenda and the 2030
Agenda containing the 17 Sustainable Development Goals for both cities in
the Global North and the Global South.[1]

As discussed previously in this volume, vital resources and opportunities
such as employment and education are often concentrated in city centres
and access to these is unevenly distributed in Europe. Access to affordable housing is one of the most pressing challenges in European cities today,
especially in the most economically vibrant ones. Consequently, access to
affordable housing is one of the core demands of the "right to the city" movement, along with other demands that focus on accessibility, such as access to

education, cultural institutions, public spaces and social networks. Since the "right to the city" is a question about equity and fairness in accessibility, it is consequently also a question about access to mobility options in the form of the "right to mobility". Verlinghieri and Venturini (2018, 127) argue:

> On one hand, the right to mobility is functional part of the right to the city, being a necessary condition for both the right to appropriation and the right to participation to be met. As such, there is an overlap between the right to mobility and the need to ensure a purposeful movement to access services, social capital, and the city [...]. The right to mobility subsumes also the right to accessibility, as fundamentally linked to questions of just access to resources and assets. On the other hand, the right to mobility goes beyond and enriches the perspective given by the right to the city bringing the attention to the key role of mobilities in the production of urban processes.

While the right to the city and the importance of access to mobility options is recognised in the development frameworks, these documents remain conceptually vague and offer little in respect of mechanisms for implementation (Uteng and Lucas 2017). This vacuum is filled to a substantial degree by a discourse around smart cities, whose proponents promise to provide technology-driven solutions to pressing urban challenges, including those of mobility and transportation. There have been frequent criticisms of these technocratic solutions, especially when they promise to combat social challenges and build "inclusive cities". On the one hand, it can be observed that technological solutions are cherished as a panacea for solving all kinds of urban challenges – including the objective of creating social cohesion – as other technology-driven approaches have previously been perceived (see e.g. Miciukiewicz and Vigar 2012; Hollands 2015; Yigitcanlar, Foth, and Kamruzzaman 2019). On the other hand, the smart cities discourse needs to be understood in the context of the global positioning of "lighthouse" or "best practice" cities. Such differentiated perspectives reveal that smart cities are less about creating equal and just cities and more about global economic competitiveness, valorisation of global city status and corporate city visions (Wiig 2015; Joss et al. 2019) and ultimately cater mostly to the interests of global business elites (Grossi and Pianezzi 2017). Although alternative visions of smart cities have been proposed, the "dominant paradigm of smart cities is still rooted in a technocratic formulation, albeit one that now acknowledges the need for citizen participation though very much from a civic paternalist or stewardship perspective" (Kitchin et al. 2017). Such a paternalistic approach is contradictory to a progressive and emancipatory approach to the "right to the city" and the "right to mobility".

As this short introduction demonstrates, an analysis of mobility poverty in urban areas is not only urgent and timely, it is also highly political. An analysis of urban inequality reveals vested interests and the concentration of power and wealth in cities and at the same time experiences of deprivation, marginalisation and exclusion. Complementing what has been written in Chapter 4, this chapter identifies some of the specifically urban dynamics that foster mobility poverty (or that are reinforced by them).

While it is clear that geography matters, it is often difficult to recognise how spatial factors contribute to social disadvantage and social exclusion in cities, and vice versa. Although there is extensive academic work on social and transport exclusion, such analysis does not usually start with spatial analysis (Dodson et al. 2006). Common spatial phenomena observed in spatial research such as residential segregation and centre-periphery relations alone cannot explain social exclusion. Therefore, some of the major urban conditions and processes will be scrutinised in their relation to mobility poverty.

Characteristics of cities in Europe in a comparative perspective

Different histories of centrally planned versus market-driven economic development have shaped cities in Europe to a large extent. Furthermore, across Europe, some cities have experienced economic and social decline, while others flourish and exhibit dynamic development. Such dynamics are almost always related to challenges for everyday mobility and accessibility problems.

In Europe, the share of the population living in cities and peri-urban areas was almost 75% by 2016 (European Union 2016). Western European countries thus show a much higher level of urbanisation than eastern member states (European Union 2016, 35–36). There are parallel and contradicting population trends in and between European regions. Unemployment and low wages due to the declining significance of heavy industries, mining and agriculture are often a push factor for young people and those of working age to leave old industrial areas and move to other cities and regions. In such cases, not only do the cities in these regions experience decline, but also the peri-urban and rural hinterlands of these cities. Demographic change further contributes to the ageing of populations in these regions.

On the other hand, there are cities that attract population not only from the immediate region, but also from the whole nation state and far beyond. These are usually the capital cities, with a high diversity of educational and cultural institutions. These cities especially attract younger and highly skilled people. Sometimes, secondary cities specialised in innovative technologies and knowledge-based economies also attract highly skilled professionals.

In addition, there are dynamics that affect certain regions of Europe in a specific way. In Ireland and Estonia, an upward economic trend made many young people decide to move to those countries, while at the same time, increasing rents in inner cities resulted in the urban outmigration of elderly people and families (Bell et al. 2010, 10–19). The incorporation of many eastern European countries into the EU has spurred labour migration from these countries to western European cities (Findlay and McCollum 2013).

Accordingly, some European countries, such as Sweden, Ireland, the Czech Republic, Finland and Spain, recorded a high population growth in cities and peri-urban regions with more than 10% population growth between 2004 and 2014.[2] Some of the largest metropolitan regions in Europe, such as London, Paris, Madrid and Rome, received between 0.5 and 1.5 million inhabitants in that period. On the other hand, there are metropolitan regions that lost population, often those centres with a historic industrial base of coal, steel and heavy industries, e.g. the Ruhr area in Germany and Katowice in Poland (European Union 2016, 66–69). These regions already face substantial challenges in maintaining a reasonable standard of public infrastructure and services, including public transportation (Reckien and Martinez-Fernandez 2011).

Differentiated urban development across Europe becomes more visible when economic development is observed. In Europe, economic activity is usually concentrated in cities to a substantial extent. It is the European capital cities and their metropolitan regions in particular that are centres for education and science and are characterised by a high social and cultural diversity. They are thus the primary centres for innovation and economic growth. Large metropolitan areas provide employment to more than 41% of the total workforce and contributed to more than 47% of the total GDP of the European Union in 2016 (European Union 2016, 61).

Access to employment and levels of income differ substantially in Europe. Income levels in European cities are usually the highest. Almost one quarter of the population of the EU-28 living in cities has an income of 150% or higher of the national median level (European Union 2016, 40). While cities create economic wealth, several western and southern European regions exhibit high levels of unemployment in urban regions (European Union 2016, 255–258). Comparing unemployment levels between different cities identifies the more dynamic and prosperous cities and regions as well as those in economic decline. Examples of dynamic and declining regions are northern France with its areas historically shaped by heavy industry and mining and the dynamic Paris metropolitan regions: unemployment rates in cities in northern France range between 21 and 23%, while unemployment rates in the Paris region are much lower (below 8%). Unemployment and limited financial resources can limit personal mobility substantially. At the same time, available transport options and individual mobility aptitudes have an impact on access to employment. Material deprivation is hence one of the crucial influencing factors of mobility poverty.

While cities in the EU are characterised by high economic growth and wealth, at the same time they exhibit a range of social inequalities. These inequalities resulting from the polarisation between economic opportunities and challenges have developed in such a way that they are often more widely observed in cities than countries as a whole.

One of these challenges in European cities is housing. The availability of affordable housing increases the attractiveness of cities. This is especially the case for low-income and other socially disadvantaged populations in search of better employment and education opportunities due to, among others, the higher availability of public transport and better accessibility. On the other hand, the low availability of affordable housing can force low-income populations to remain in rural or peri-urban areas or move out of cities where usually fewer public transport options are available and there is lower accessibility to employment.

The housing costs in cities of western Europe are considerably higher than in its rural areas, while in some eastern states (Bulgaria, Slovakia, Croatia, Romania) the housing costs in rural areas are higher than in cities. Furthermore, in capital cities of western and northern Europe, the cost of living is the highest in the European Union. On top of the list is London, followed by Copenhagen, Stockholm, Helsinki, Dublin and Paris. Among these cities, London and Stockholm experienced a substantial rise in living costs between 2005 and 2015 (European Union 2017).

Urban deprivation in Europe

The challenges encountered in many cities are multi-fold. An increase in poverty, low levels of education, lack of investment in child and youth development and increasing gaps between rich and poor citizens are not only phenomena observed in cities, but they often appear in cities in a concentrated form. Poverty tends to cluster in certain urban neighbourhoods. Furthermore, poor citizens live at an increasing distance from wealthier citizens. Varying living conditions in different parts of a city can potentially have negative effects on social mobility, since the quality of schools, access to services and decent living conditions are important for people to prosper and fulfil their potential. For wider society, social and economic polarisation can be severe for the social fabric.

European cities are historically characterised by social diversity. However, increasing income divides, decreasing security of employment, deregulation of housing markets and shrinking welfare states are some of the economic changes that have contributed to increasing social and economic polarisation in cities. In many western European countries, the suburbanisation of wealthier residents occurred – also fostered by tax incentives as in the case of Germany (Rohrbach 2003) – while low-income residents often remained in the urban core. Due to migration within European states and from outside Europe, parts of cities have become characterised by a strong ethnic concentration (Colini et al. 2013, 8). More recently, due to gentrification and tight rental housing markets in inner cities, low-income residents are compelled to move to peripheral locations or even low-density suburban sites. In low-density suburban areas, public transport is usually less frequently available (Korsu and Wenglenski 2010). The analysis of spatial segregation within cities often reveals the socio-economic (or ethnic and religious) divides that may ultimately lead to a situation where whole urban districts embark on a downward spiral of deprivation. When situations of social disadvantage meet negative "site effects" of deprived urban neighbourhoods, these conditions can mutually reinforce each other. The effect on the mobility situation may be twofold: either frequent travel of residents is necessary to access work, education, health services, etc. but is not provided – or not at sufficient levels – at the place of residence or mobility is impeded due to limited or low-quality mobility options available in these areas (O'Connor, Borscheid, and Reid 2013).

Spatial segregation in Europe

The observation of segregation is a methodology to identify social divides in cities. However, it can only observe the spatial concentration of the population with specific indicators, while not being able to make statements about the actual living situation of residents; such observation can neither reveal how residents deal with their (disadvantaged or privileged) situation, nor whether individuals or households are actually negatively affected by negative location effects. Hence, an analysis of segregation is only the first step towards detecting actual disadvantage and marginalisation and, ultimately, mobility poverty.

Spatial segregation along social, economic and ethnic lines is increasing throughout European cities. However, the pattern of segregation differs across Europe. In Denmark and the Netherlands, for example, the poorest households show the highest level of spatial concentration, while in France it is the most affluent who tend to concentrate in specific areas of a city. Hence, in these countries, segregation is relatively more driven by the most affluent than by the poor (see Figure 5.1). This development draws attention to the functioning of the housing sector. Land-use regulations may have exclusionary effects for low-income households in certain neighbourhoods (Arbaci 2007, 420–422). The rise of private communities – condominiums, housing co-operatives and "gated communities" – may also have contributed to the segregation of the wealthier citizens within cities and metropolitan areas (Musterd 2017, 251–253).

Generally, there is an increase in the level of segregation in European cities. Levels of socio-economic segregation were higher in 2011 on average than those in 2001. This is evident e.g. when observing 12 European capital cities by income, type of occupation and educational attainment using a Dissimilarity Index as a measure of segregation (see Figure 5.2). Socio-economic segregation has increased in all of the capital cities analysed except Amsterdam. Among the cities considered, Madrid exhibited the highest level of segregation in 2011. Madrid was closely followed by Tallinn and London. Madrid and Tallinn also had the sharpest increase in socio-economic segregation between 2001 and 2011, together with Stockholm.

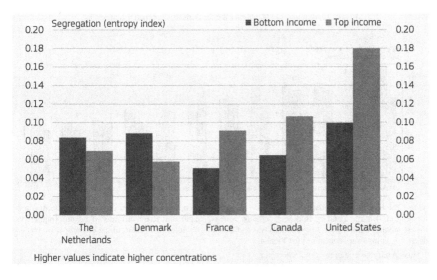

Figure 5.1 Income concentration in cities by income group, 2017.
Source: European Commission 2017, 78, based on OECD 2016.

The lowest level of segregation among these cities was observed in Oslo, followed by Riga and Prague (OECD 2016, 75–81).

It needs to be noted that, while spatial segregation is commonly acknowledged as an indicator of a social divide at urban level, social diversity in urban neighbourhoods is not necessarily associated with social cohesion and harmony, as social groups can live alongside each other with little or no interaction (Shaw and Hagemans 2015; Boterman and Musterd 2016). Similarly, ethnic neighbourhood diversity is neither sufficient for social cohesion and cultural interchange, nor are multi-ethnicity and multi-cultural factors in urban neighbourhoods explanations for ethnic intolerance and social conflicts (Amin 2002, 960). As often the case, the reasons for urban deprivation and social inequalities are complex and multi-layered and so are the potential strategies and solutions.

It has been argued that state and business-driven development and urban regeneration have led to selective accessibility in wealthier parts of the city, by deploying premium – sometimes private – infrastructures such as toll highways and modern rail systems. Such developments contribute to the socio-spatial fragmentation of cities and relationally disadvantage the less affluent population concentrated in deprived urban areas. Such elite enclaves are often better connected globally through physical and virtual networks than embedded in the local urban fabric (Graham and Marvin 2001; Amin 2013; Wissink, Schwanen, and van Kempen 2016). On the other hand, for poor households in large European cities, an increasing spatial mismatch of housing and job locations can be observed. In some

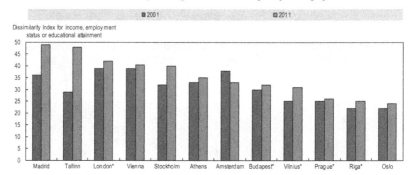

Index of Dissimilarity: The higher the index the higher spatial segregation

Notes: The Index of Dissimilarity was computed in terms of occupation (managers vs. elementary occupations) for Madrid, Tallinn, London, Budapest, Vilnius, Athens, Prague and Riga; in terms of income (highest vs. lowest income quintile) for Amsterdam, Oslo and Stockholm; in terms of educational attainment (university degree vs. compulsory education) for Vienna.

* Municipality instead of metropolitan region.

Figure 5.2 Change in spatial segregation of major European cities, 2001–2011.
Source: OECD 2016, 76. Reprinted by permission from OECD Publishing: Making Cities Work for All: Data and Actions for Inclusive Growth by OECD © 2016.

cases – such as Paris – an efficient public transport system allows sufficient access to such jobs. However, in other cases such as London where public transport provision is more deregulated and market driven, service levels of public transport are not equal across the metropolitan region (Coutard, Dupuy, and Fol 2004).

Deprivation on large housing estates

The spatial concentration of poverty and social disadvantages has also been driven by planning interventions in the 1950s–1980s, when large mono-functional housing estates were built in eastern and western European states to meet the needs of the growing urban population (Power 2012). While many of these estates were well equipped with facilities according to modern standards, others were built in a very short time, with inadequate infrastructure and facilities for social, education and supply services. The popularity of those estates declined and were used by housing officials to rehouse disadvantaged groups, including, among others, migrants to the city. Since the 1990s, these neighbourhoods have been seen as the most problematic areas of many cities, with problems ranging from physical downgrading of the housing stock, increasing concentrations of low-income households, rising criminality and decreasing quality of public space. Since then, across Europe, these areas have been targeted by a range of policy strategies (Wassenberg 2004; van Kempen, Wassenberg, and van Meer 2007; Dekker et al. 2011).

The socio-demographic and economic situation of large housing estates differs in different European regions. In eastern Europe, these estates are still able to attract the middle classes, especially well-educated families, artists and young childless couples. Hence, income levels are higher and unemployment levels are lower than in the rest of the cities in eastern Europe. The situation in southern, western and northern Europe is different. Especially in southern Europe, the original population of large housing estates is ageing, leading to an overrepresentation of the elderly. As the original population is ageing and diminishing, migrant end ethnic minority families are moving into relatively spacious apartments. This can be observed particularly in western and northern Europe. In western Europe, on some of these estates, over 80% of the total population belongs to ethnic minority groups, usually with low incomes and few opportunities on the urban housing market.

Although the presence of a mix of elderly and migrant/ethnic minority populations in these areas has produced social-cultural heterogeneity, this is often perceived negatively, especially by those who have been long established in these neighbourhoods. At least partially, the over-presence of ageing and migrant populations gives way to unemployment levels which are usually higher on large housing estates in southern, western and northern member countries. Employment opportunities for ethnic minority members and migrants are often lower; young people with a migrant background in particular experience difficulties in finding jobs. Another reason is that the initial design of these estates did not offer space for production and services, but was foremost oriented towards providing residential space (Dekker and van Kempen 2004, 573–574).

Today, the physical condition of these large-scale housing estates is a factor that aggravates the social situation in these neighbourhoods. Maintenance is often problematic, especially on those built several decades ago and where low-quality building material was used, resulting in many physical problems in this building stock. Despite often being located between large tracts of greenery, they are often perceived as grey and monotonous (Dekker and van Kempen 2004, 572–573).

Residents living in deprived urban districts and housing estates often have to deal with the stigma associated with their place of residence. Pierre Bourdieu and his colleagues has famously argued that the urban is a reification of social space, and that "inhabited [...] space function[s] as [...] spontaneous symbolization of social space" (Bourdieu, Accardo, and Emanuel 1999, 124). They argued that negative symbolic meaning in the form of stigmatisation "intensifies the experience of finitude: it chains one to a place". This observation inspired a whole stream of research on the negative reputation of urban districts (e.g. Garbin and Millington 2011; Slater 2017). Most notable are the works of Loïc Wacquant on deprived urban districts and housing estates in cities of the United States and France. He argued that territorial stigmatisation is "arguably the single most protrusive feature of the lived experience of those trapped in these sulphurous zones" (Wacquant 2008, 169).

Do large housing estates with the abovementioned characteristics foster mobility poverty or, conversely, does mobility poverty create or foster social exclusion on such estates? While such an association can be drawn intuitively, the relationship is not always evident in empirical findings. Anne Power (2012) has observed mobility disadvantages and housing estates in deprived conditions with a concentration of poverty in the United Kingdom, while at the same time she recognised that such estates are also over-proportionally affected by the externalities of car traffic, in the form of noise and pollution. A study from a low-income area with housing estates in northern Dublin highlighted the perception of the poor service quality of public transport in the area. It is argued that the negative perception of the services increases the potential for social marginalisation and diminishes self-esteem among residents (O'Connor, Borscheid, and Reid 2013).

The depreciation of large housing estates and experiences of mobility disadvantages are also not a universal European phenomenon because they are largely confined to western and southern Europe. However, not everywhere in western Europe do large social housing estates experience deprivation and stigmatisation either. As evidence from Austria shows, some of the large public housing estates are preferred places of residence and are well served with mobility options (Oberzaucher 2017; Zupan 2020). In eastern Europe, however, large housing estates have never been stigmatised as a whole; many middle-income households live in such large housing areas (Bolt 2018). Surveys from eastern Germany and the Czech Republic indicate that housing estates are still considered a favourable form of residence (Grossmann, Kabisch, and Kabisch 2015). Large housing estates in eastern Europe today benefit from the fact that the development of public transport on such estates was given priority, with the result that access to mobility options may be even better in such areas of cities (Leetmaa et al. 2018).

Summarising the findings, it can be stated that living on large housing estates does not necessarily mean that mobility disadvantage is experienced or that social exclusion is fostered by mobility poverty. Nevertheless, in situations where deprivation of housing estates and concentration of poverty is evident, access to mobility options can make an enormous difference to residents. If sufficient high-quality options are available, it can contribute positively to people's living situation in substantial ways.

Transformation of declining inner-city neighbourhoods

Large housing estates are not the only areas where deprivation and processes of decline can be observed. Historically, many inner-city neighbourhoods, especially the city extensions of the late 19th and early 20th century, were established to accommodate the working classes of growing urban populations. In the second half of the 20th century, in many European cities, migrants replaced the aging population in these neighbourhoods, living alongside low-income domestic residents. Like large housing estates, some of these neighbourhoods have experienced processes of urban decay, economic decline and stigmatisation (see e.g. Breckner 2013 for Hamburg/ Germany). However, although these districts have been decaying, migrants and low-income residents benefited from well-developed public transport and convenient access to locations of employment, education and health services.

These inner-city neighbourhoods have been transformed in recent decades, pushing the former low-income and migrant populations to the edges or out of the city, resulting in new mobility and accessibility challenges for these groups. This process has been driven by a partial reorientation in residential choices of middle- and high-income populations, who more often choose inner city neighbourhoods as their preferred places of residence. This tendency is a counter-dynamic to the dominance of suburbanisation in previous decades and contributes to tense urban housing markets (Hierse et al. 2017).

Many authors have pointed to the benefits of the "return" of the middle classes into formerly decaying inner city areas. The presence of the middle classes usually leads to landscape and infrastructure upgrades and attracts public investment to the area. Often the benefits of improvement in education and job prospects for poorer residents are highlighted, resulting in the opportunity of upward social mobility. However, it has been observed that the process of gentrification leads to higher rents in the respective neighbourhoods (Shaw and Hagemans 2015; Marcińczak et al. 2016; Alexandri 2018). As more inner city areas are regenerated and upgraded, low-income residents are forced to move to the peripheries of cities, or even to suburban locations (Hochstenbach and Musterd 2018). Processes of gentrification, dynamic upgrading of inner-city areas and replacement of low-income residents by urban middle classes have been described most extensively for western European cities. However, similar processes are observed in the post-socialist cities of today's eastern European member states, such as Budapest, Vilnius and Prague (Brade, Herfert, and Wiest 2009; Fabula et al. 2017).

Peripheral urban areas: the case of Roma communities

Finally, the condition of peripherality in relation to inner cities and the consequences for everyday mobility must be observed. Inadequate access to employment, education, supply infrastructure and other institutions cannot generally be associated with peripheral locations. Such opportunities may be equally well developed outside the inner city or well-developed public transport can offset latent disadvantages of urban location. However, when deprivation appears in peripheral housing locations, coupled with few employment opportunities, access to services and insufficient public transport, peripherality can mean a severe social disadvantage and can lead to mobility poverty (see e.g. Murie and Musterd 2004). In this regard, the situation of a minority group that experiences the most severe forms of deprivation, material poverty and social exclusion in Europe needs to be addressed. The Roma are Europe's largest and most vulnerable minority. In total, about 7 million to 9 million Roma live in Europe (Ringold, Orenstein, and Wilkens 2005, xii–xic, 155–158).[3] Forms of living for the Roma differ from country to country, as does the proportion living in cities or rural settlements. While many Roma live integrated in urban neighbourhoods, marginal and segregated settlements of Sinti and Roma can be found all over European cities. While this is often a result of limited access to housing markets, some Roma communities have chosen to live separately or hope to avoid barriers of discrimination.

Throughout the European Union, particularly in south-eastern Europe, Roma communities living in segregated settlements are experiencing conditions of extreme poverty and social marginalisation. Alongside education, employment and health care, housing and settlement issues are among the most difficult and pressing challenges of the Sinti and Roma population in Europe. Segregated Roma settlements across the European Union are characterised by substandard living conditions and insecure residence, coupled with the threat of forced evictions, lack of civil registration and inability to access employment and education opportunities (Ringold, Orenstein, and Wilkens 2005; OSCE 2006).

There is an evidence-based link between the geographic location of Roma settlements, housing conditions and the poverty of the communities. The housing policies of former and recent governments have often led to the regional and geographic isolation and segregation of Roma neighbourhoods. These settlements are found in peripheral and disadvantaged urban locations, often in hazardous areas such as next to or between highways and railway lines or environmentally degraded and contaminated areas. Formal basic infrastructure provision is either inadequate or absent and settlements often located outside the public transport catchment area, with the result that Roma communities have limited access to city centres with their public services, employment and education opportunities (Ringold, Orenstein, and Wilkens 2005, 34–38, 94; Bermann and Clough Marinaro 2014, 409). Marinaro observed that a Roma settlement on the outskirts of Rome was entirely isolated from other residential areas as well as shops and other services. From the settlement, it takes roughly two hours to reach Rome city centre by public transport and the nearest bus stop is one and a half kilometres away. Residents further state that buses rarely stop to pick them up (Marinaro 2009, 278–279). Poor public transport is also frequently mentioned as a barrier for children to attend school (Ringold, Orenstein, and Wilkens 2005, 129–135; European Parliament 2008, 147; Di Giovanni 2014, 7).

Conclusion

It has become apparent that mobility is a major challenge in urban areas. Insufficient transport options are often characteristic of large housing estates and peripheral deprived areas. This impediment to personal mobility thus further deprives residents living in these areas who often already experience forms of social disadvantages. The situation is further aggravated when certain neighbourhoods in cities are faced with stigmatisation.

Examples show that the improvement of transport services can break the vicious cycle of deprivation and marginalisation. However, more research is needed to fully understand the impact of mobility poverty on urban deprived areas. Certainly, the challenges need to be addressed by carefully developed, integrated solutions that do not only address the mobility situation, but also the housing situation and provision of other basic services. Also, the possible outcomes of gentrification processes need to be taken into account when drafting solutions in order not to create negative effects for low-income and vulnerable groups in these areas.

Notes

1 The New Urban Agenda highlights that inclusive urban public spaces are key drivers for social and economic development (article 53) and that the design of urban and metropolitan transport schemes is related, among others, to social cohesion, quality of life and accessibility (article 115) (United Nations 2017). Hence, although mobility poverty in urban areas is usually not employed as a term, experiences of mobility disadvantage take centre stage in planning and development at all levels today.
2 These figures are reported at regional level (NUTS 3), using the urban-rural typology, instead of local administrative level (LAU), using the degree of urbanisation typology.
3 Estimates of the number of Roma in Europe differ widely. The share of Roma in Bulgaria, Slovakia and Romania is estimated at 6% to 9% of the population. Romania has the highest absolute number of Roma in Europe, with between 1 million and 2 million. Large populations of between 400,000 and 1 million also live in Bulgaria, Hungary and Slovakia. Western Europe's largest Roma populations are found in Spain (estimated at 630,000), France (310,000), Italy (130,000) and Germany (70,000) (Ringold, Orenstein, and Wilkens 2005).

References

Alexandri, Georgia. 2018. "Planning gentrification and the 'absent' state in Athens." In *International Journal of Urban and Regional Research* 42 (1): 36–50. https://doi.org/10.1111/1468-2427.12566.

Amin, Ash. 2002. "Ethnicity and the multicultural city: Living with diversity." In *Environment and Planning A* 34 (6): 959–980. https://doi.org/10.1068/a3537.

Amin, Ash. 2013. "Telescopic urbanism and the poor." In *City* 17 (4): 476–492. https://doi.org/10.1080/13604813.2013.812350.

Arbaci, Sonia. 2007. "Ethnic segregation, housing systems and welfare regimes in Europe." In *International Journal of Housing Policy* 7 (4): 401–433. https://doi.org/10.1080/14616710701650443.

Bell, Simon, Susana Alves, Eva Silveirinha de Oliveira, and Affonso Zuin. 2010. "Migration and land use change in Europe: A review." In *Living Reviews in Landscape Research* 4. https://doi/10.12942/lrlr-2010-2.

Bermann, Karen, and Isabella Clough Marinaro. 2014. "'We work it out': Roma settlements in Rome and the limits of do-it-yourself." In *Journal of Urbanism: International Research on Placemaking and Urban Sustainability* 7 (4): 399–413. https://doi.org/10.1080/17549175.2014.952321.

Bolt, Gideon. 2018. "Who is to blame for the decline of large housing estates? An exploration of socio-demographic and ethnic change." In *Housing Estates in Europe: Poverty, Ethnic Segregation and Policy Challenges*, edited by Daniel Baldwin Hess, Tiit Tammaru, and Maarten van Ham, 57–74. Cham, Switzerland: Springer.

Boterman, Willem R., and Sako Musterd. 2016. "Cocooning urban life: Exposure to diversity in neighbourhoods, workplaces and transport." In *Cities* 59: 139–147. https://doi.org/10.1016/j.cities.2015.10.018.

Bourdieu, Pierre, Alain Accardo, and Susan Emanuel. 1999. *The Weight of the World: Social Suffering in Contemporary Society*. Oxford: Polity Press.

Brade, Isolde, Günter Herfert, and Karin Wiest. 2009. "Recent trends and future prospects of socio-spatial differentiation in urban regions of Central and Eastern Europe: A lull before the storm?" In *Cities* 26 (5): 233–244. https://doi.org/10.1016/j.cities.2009.05.001.

Breckner, Ingrid. 2013. "Urban poverty and gentrification: A comparative view on different areas in Hamburg." In *Spaces of the Poor: Perspectives of Cultural Sciences on Urban Slum Areas and Their Inhabitants*. Vol. 17, edited by Hans-Christian Petersen, 193–208. Mainz: Transcript Verlag.

Colini, Laura, Darinka Czischke, Simon Guntner, Ivan Tosics, and Peter Ramsden. 2013."Against divided cities in Europe." https://www.enhr.net/documents/Papers%20Spain/Papers/WS25/Colini-Czischke-GuntnerTosics-Ramsden%20ENHR%202013.pdf. Accessed 11 June 2018.

Coutard, Olivier, Gabriel Dupuy, and Sylvie Fol. 2004. "Mobility of the poor in two European metropolises: Car dependence versus locality dependence." In *Built Environment* 30 (2): 138–145. https://doi.org/10.2148/benv.30.2.138.54313.

Dekker, Karien, and Ronald van Kempen. 2004. "Large housing estates in Europe: Current situation and developments." In *Tijdschrift voor economische en sociale geografie* 95 (5): 570–577. https://doi.org/10.1111/j.0040-747X.2004.00340.x.

Dekker, Karien, Sjoerd de Vos, Sako Musterd, and Ronald van Kempen. 2011. "Residential satisfaction in housing estates in European cities: A multi-level research approach." In *Housing Studies* 26 (04): 479–499. https://doi.org/10.1080/02673037.2011.559751.

Di Giovanni, Elisabetta. 2014. "Ethnically unprivileged: Some anthropological reflections on roma women in Contemporary Italy." In *Facets of Women's Migration*, edited by Elisabetta Di Giovanni, 5–13. Newcastle upon Tyne: Cambridge Scholars Publishing.

Dodson, Jago, Nick Buchanan, Brendan Gleeson, and Neil Sipe. 2006. "Investigating the social dimensions of transport disadvantage—I. Towards new concepts and methods." In *Urban Policy and Research* 24 (4): 433–453. https://doi.org/10.1080/08111140601035317.

European Commission. 2017. *My Region, My Europe, Our Future: The Seventh Report on Economic, Social and Territorial Cohesion*. Report on economic, social and territorial cohesion 7. Luxembourg: Publications Office of the European Union.

https://ec.europa.eu/regional_policy/en/information/publications/communications/ 2017/myregion-my-europe-our-future-the-seventh-report-on-economic-social- and-territorial-cohesion. Accessed 11 June 2018.

European Parliament. 2008. "The social situation of the Roma and their improved access to the labour market in the EU." www.europarl.europa.eu/RegData/ etudes/etudes/join/2008/408582/IPOL-EMPL_ET(2008)408582_EN.pdf. Accessed 12 June 2018.

European Union. 2016. "Urban Europe: Statistics on cities, towns and suburbs." https://ec.europa.eu/eurostat/statistics-explained/index.php/Urban_Europe_% E2%80%94_statistics_on_cities,_towns_and_suburbs. Accessed 13 June 2018.

European Union. 2017. "Eurostat regional yearbook: 2017 edition." https://ec. europa.eu/eurostat/documents/3217494/8222062/KS-HA-17-001-EN-N.pdf/eaebe 7fa-0c80-45af-ab41-0f806c433763. Accessed 13 June 2018.

Fabula, Szabolcs, Lajos Boros, Zoltán Kovács, Dániel Horváth, and Viktor Pál. 2017. "Studentification, diversity and social cohesion in post-socialist Buda- pest." In *Hungarian Geographical Bulletin* 66 (2): 157–173. https://doi.org/10.15201/ hungeobull.66.2.5.

Findlay, Allan, and David McCollum. 2013. "Recruitment and employment re- gimes: Migrant labour channels in the UK's rural agribusiness sector, from acces- sion to recession." In *Journal of Rural Studies* 30: 10–19. https://doi.org/10.1016/j. jrurstud.2012.11.006.

Garbin, David, and Gareth Millington. 2011. "Territorial stigma and the politics of resistance in a Parisian banlieue: La Courneuve and beyond." In *Urban Studies* 49 (10): 2067–2083. https://doi.org/10.1177/0042098011422572.

Graham, Stephen, and Simon Marvin. 2001. *Splintering Urbanism: Networked Infra- structures, Technological Mobilities and the Urban Condition*. London: Routledge.

Grossi, Giuseppe, and Daniela Pianezzi. 2017. "Smart cities: Utopia or neoliberal ideology?" In *Cities* 69: 79–85. https://doi.org/10.1016/j.cities.2017.07.012.

Grossmann, Katrin, Nadja Kabisch, and Sigrun Kabisch. 2015. "Understanding the social development of a post-socialist large housing estate: The case of Leipzig- Grünau in eastern Germany in long-term perspective." In *European Urban and Regional Studies* 24 (2): 142–161. https://doi.org/10.1177/0969776415606492.

Hierse, Lin, Henning Nuissl, Fabian Beran, and Felix Czarnetzki. 2017. "Concur- ring urbanizations? Understanding the simultaneity of sub-and re-urbanization trends with the help of migration figures in Berlin." In *Regional Studies, Regional Science* 4 (1): 189–201. https://doi.org/10.1080/21681376.2017.1351886.

Hochstenbach, Cody, and Sako Musterd. 2018. "Gentrification and the suburbani- zation of poverty: Changing urban geographies through boom and bust periods." In *Urban Geography* 39 (1): 26–53. https://doi.org/10.1080/02723638.2016.1276718.

Hollands, Robert G. 2015. "Critical interventions into the corporate smart city." In *Cambridge Journal of Regions, Economy and Society* 8 (1): 61–77. https://doi. org/10.1093/cjres/rsu011.

Joss, Simon, Frans Sengers, Daan Schraven, Federico Caprotti, and Youri Dayot. 2019. "The smart city as global discourse: Storylines and critical junctures across 27 cities." In *Journal of Urban Technology* 26 (1): 3–34. https://doi.org/10.1080/106 30732.2018.1558387.

Kitchin, Rob, Claudio Coletta, Leighton Evans, Liam Heaphy, and Darach Mac Donncha. 2017. "Smart cities, urban technocrats, epistemic communities and ad- vocacy coalitions." The Programmable City Working Paper 26. Prepared for the

'A New Technocracy' workshop, University of Amsterdam, 20–21 March 2017. http://progcity.maynoothuniversity.ie/. Accessed 25 March 2020.

Korsu, Emre, and Sandrine Wenglenski. 2010. "Job accessibility, residential segregation and risk of long-term unemployment in the Paris region." In *Urban Studies* 47 (11): 2279–2324. https://doi.org/10.1177/0042098009357962.

Leetmaa, Kadri, Johanna Holvandus, Kadi Mägi, and Anneli Kährik. 2018. "Population shifts and urban policies in housing estates of Tallinn, Estonia." In *Housing Estates in Europe: Poverty, Ethnic Segregation and Policy Challenges*, edited by Daniel Baldwin Hess, Tiit Tammaru, and Maarten van Ham, 389–412. Cham, Switzerland: Springer.

Marcińczak, Szymon, Sako Musterd, Maarten van Ham, and Tiit Tammaru. 2016. "Inequality and rising levels of socio-economic segregation: Lessons from a pan-European comparative study." In *Socio-economic Segregation in European Capital Cities: East Meets West*, edited by Szymon Marcińczak, Sako Musterd, Tiit Tammaru, and Maarten van Ham, 358–382. London: Routledge.

Marinaro, Isabella C. 2009. "Between surveillance and exile: Biopolitics and the Roma in Italy." In *Bulletin of Italian Politics* 1 (2): 265–287. https://www.gla.ac.uk/media/Media_140172_smxx.pdf. Accessed 8 April 2020.

Miciukiewicz, Konrad, and Geoff Vigar. 2012. "Mobility and social cohesion in the splintered city: Challenging technocentric transport research and policy-making practices." In *Urban Studies* 49 (9): 1941–1957. https://doi.org/10.1177/0042098012444886.

Murie, Alan, and Sako Musterd. 2004. "Social exclusion and opportunity structures in European cities and neighbourhoods." In *Urban Studies* 41 (8): 1441–1459. https://doi.org/10.1080/0042098042000226948.

Musterd, Sako. 2017. "Growing socio-spatial segregation in European capitals: Different government, less mitigation." In *Urban Europe. Fifty Tales of the City*, edited by Virginie Mamadouh and A. Wageningen, 251–259. Amsterdam: Amsterdam University Press.

Oberzaucher, Elisabeth. 2017. "Der Wiener Gemeindebau als Vorbild für den sozialen Wohnungsbau." In *Homo urbanus: Ein evolutionsbiologischer Blick in die Zukunft der Städte*, edited by Elisabeth Oberzaucher, 175–186. Berlin and Heidelberg: Springer.

O'Connor, David, Matthias Borscheid, and Odran Reid. 2013. "An assessment of mobility among key disadvantaged communities in North East Dublin." In *Proceedings of the AESOP-ACSP Joint Congress, Dublin 2013*. https://arrow.tudublin.ie/cgi/viewcontent.cgi?article=1014&context=beschspcon. Accessed 8 April 2020.

OECD. 2016. "Making cities work for all." OECD Publishing, Paris. https://doi.org/10.1787/9789264263260-en. Accessed 11 June 2018.

OSCE. 2006. "Roma housing and settlements in South-Eastern Europe: Profile and achievements in Serbia in a comparative framework." https://www.osce.org/odihr/23336?download=true. Accessed 12 June 2018.

Power, Anne. 2012. "Social inequality, disadvantaged neighbourhoods and transport deprivation: An assessment of the historical influence of housing policies." In *Journal of Transport Geography* 21: 39–48. https://doi.org/10.1016/j.jtrangeo.2012.01.016.

Reckien, Diana, and Cristina Martinez-Fernandez. 2011. "Why do cities shrink?" In *European Planning Studies* 19 (8): 1375–1397. https://doi.org/10.1080/09654313.2011.593333.

Ringold, Dena, Mitchell A. Orenstein, and Erika Wilkens. 2005. *Roma in an Expanding Europe: Breaking the Poverty Cycle:* Washington, DC: World Bank. http://documents.worldbank.org/curated/en/600541468771052774/pdf/30992.pdf. Accessed 27 March 2020.

Rohrbach, Iris. 2003. "Eigenheimzulage–Ex-post-Analysen zu ausgewählten Reformvorschlägen." In *Informationen zur Raumentwicklung* 10 (6): 355–365. https://www.bbsr.bund.de/BBSR/DE/Veroeffentlichungen/IzR/2003/Downloads/6Rohrbach.pdf?__blob=publicationFile&v=2. Accessed 8 April 2020.

Shaw, Kate S., and Iris W. Hagemans. 2015. "'Gentrification without displacement' and the consequent loss of place: The effects of class transition on low-income residents of secure housing in gentrifying areas." In *International Journal of Urban and Regional Research* 39 (2): 323–341. https://doi.org/10.1111/1468-2427.12164.

Slater, Tom. 2017. "Territorial stigmatization: Symbolic defamation and the contemporary metropolis." In *The SAGE Handbook of New Urban Studies*, edited by John A. A. Hannigan and Greg Richards, 111–125. London: Sage Publications.

United Nations. 2017. "New urban agenda." http://habitat3.org/wp-content/uploads/NUA-English.pdf. Accessed 25 March 2020.

Uteng, Tanu Priya, and Karen Lucas. 2017. "The trajectories of urban mobilities in the Global South: An introduction." In *Urban Mobilities in the Global South*, edited by Tanu Priya Uteng and Karen Lucas, 1–18. London: Routledge.

van Kempen, Ronald, Frank Wassenberg, and Annelien van Meer. 2007. "Upgrading the physical environment in deprived urban areas: Lessons from integrated policies." In *Informationen zur Raumentwicklung* (8): 487–497. https://www.bbsr.bund.de/BBSR/EN/Publications/IzR/2007/7_8KempenWasserbergMeer.pdf?__blob=publicationFile&v=2. Accessed 8 April 2020.

Verlinghieri, Ersilia, and Federico Venturini. 2018. "Exploring the right to mobility through the 2013 mobilizations in Rio de Janeiro." In *Journal of Transport Geography* 67: 126–136. https://doi.org/10.1016/j.jtrangeo.2017.09.008.

Wacquant, Loïc. 2008. *Urban Outcasts: A Comparative Sociology of Advanced Marginality.* Cambridge and Malden, MA: Polity Press.

Wassenberg, Frank. 2004. "Large social housing estates: From stigma to demolition?" In *Journal of Housing and the Built Environment* 19 (3): 223–232. https://doi.org/10.1007/s10901-004-0691-2.

Wiig, Alan. 2015. "IBM's smart city as techno-utopian policy mobility." In *City* 19 (2–3): 258–273. https://doi.org/10.1080/13604813.2015.1016275.

Wissink, Bart, Tim Schwanen, and Ronald van Kempen. 2016. Beyond residential segregation: Introduction. In *Cities* 59: 126–130. https://doi.org/10.1016/j.cities.2016.08.010.

Yigitcanlar, Tan, Marcus Foth, and Liton Kamruzzaman. 2019. "Towards post-anthropocentric cities: Reconceptualizing smart cities to evade urban ecocide." In *Journal of Urban Technology* 26 (2): 147–152. https://doi.org/10.1080/10630732.2018.1524249.

Zupan, Daniela. 2020. "De-constructing crisis: Post-war modernist housing estates in West Germany and Austria." In *Housing Studies*, 1–25. https://doi.org/10.1080/02673037.2020.1720613.

6 The rural arena

Stefano Borgato, Silvia Maffii and Cosimo Chiffi

Abstract

From a transportation point of view, rural areas have always represented a challenge. Their low-density population, distance from built-up areas and railway stations, shrinking and ageing population, decentralised services are among the multiple factors that hinder the provision of adequate and efficient transport services. The scenario is even more complex when vulnerable groups not capable of moving independently are involved. Many innovative transport solutions are available, but almost all the time they stay within the urban context, outside of which they fail to reach the necessary critical mass. Given this scenario, this chapter analyses the mobility-related characteristics of rural areas before going deeper into the critical challenges related to (in)accessibility and social exclusion linked to persistent mobility poverty conditions.

Introduction

As mentioned in the introductory chapter, it is difficult to establish a single definition for the urban and this applies in the same way to the "rural". Due to a process of urban and rural restructuring, it is increasingly difficult to differentiate between urban and rural economies, societies and lifestyles. Different academic disciplines, even in social sciences, have established different concepts and definitions of rurality, some more positivistic, but increasingly approaching rurality from alternative and progressive theoretical trajectories. This conceptual plurality also means that a single and universal definition of "rural mobilities" and mobility poverty in rural areas is not straightforward. While identifying the dynamics of change in rural areas is key to making sense of contemporary processes of restructuration of the countryside, it is also necessary to analyse how people and institutions socially construct "rurality" and how concepts of rurality take effect on actual restructuring of rural landscapes (Woods 2005, 299). What we propose here is to understand and analyse rural processes and diversity in such a double context and not to separate them due to their different philosophical start-ing points. Rural space in a Lefebvrian sense is socially produced by imag-inative, material and practised ruralities that form the rural as a totality (Woods 2009, 851–852; see also Halfacree 2006). Following such a perspec-tive requires not only the analysis of mobilities in and between rural areas (and between rural and urban areas), but also investigating how mobilities, on the one hand, reshape rural places and, on the other hand, (re)produce rurality. Such an approach opens up rich pathways for a comprehensive un-derstanding of mobility poverty in rural areas.

The rural has been, and still is to some degree, approached in academia as a functional concept, most often in its opposition to the urban and hence describing the rural-urban relationship. According to such a functional conceptualisation, rural areas are dominated by specific extensive land uses such as agriculture and forestry, are spatially structured by small and dis-persed settlements and characterised by ways of living based on cohesive identities linked to land use and settlement structure (Cloke 2006, 20). Policy advice and political decision-making is still guided – to a large degree – by these functional concepts, due to the application of technology and its need for quantitative methods, but also due to the political will to respond to rural needs with policy programmes, requiring new "objective" rural classi-fications and sets of indicators (see e.g. Cloke 2006, 20). There needs to be a clear distinction between the political and developmental potentials of such rural concepts, and the analytical benefit for research of these concepts; as a result of that careful distinction, research needs to clearly highlight the shortcomings of policies based on such functional spatial differentiations, especially when it comes to an intrinsically fluid and multi-layered topic like mobilities and mobility poverty.

Rural areas in Europe have undergone considerable restructuring. Once densely populated spaces dominated by agriculture and other primary-sector activities are today reduced in population density and marked by a variety of production activities and service economies. Rural areas have also become places of consumption: some regions that heavily rely on tour-ism are even overwhelmingly dominated by consumption (Halfacree 2006, Silva and Figueiredo 2013, Woods 2005, 2011). Economic restructuring and urbanisation have led to a general loss in rural population, although pro-cesses of counter-urbanisation have been described in several European countries. Structural shifts and processes of decline have been particularly dramatic in remote and peripheral rural areas (Silva and Figueiredo 2013). Most of these dynamics are related in one way or another to processes of "modernisation", such as technological innovation and organisational changes in agriculture, industries and everyday life. Many of these tech-nological advances have enabled people to become more mobile, physi-cally and virtually, changing people's relationship to their place of living (Woods 2005).

Political economists and geographers have analysed the driving factors of continuous rural restructuring. They highlight how rural areas are increasingly characterised by their connections to the regional, national and global scale, making apparent that "much of what happens within rural areas is caused by factors operating outside the supposed boundaries of these areas" (Cloke 2006, 20). Such factors relate to economic globalisation that result in traditional rural agricultural products being traded on global markets that are dominated by a few transnational corporations (Woods 2005). Foreign labour migrants work as seasonal labourers in the agricultural sector and state authorities accommodate refugees in rural areas, resulting in an increasingly heterogeneous social composition of rural areas (Bell and Osti 2010, Smith 2007). All of these dynamics hint towards intensified and diversified circulation between rural and urban areas, but also between different rural areas (Hedberg and do Carmo 2012). In response to these dynamics, rural areas have been the focus of state responses such as policy reforms, land use and trade regulation, environmental protection and so on. In the face of rural deprivation, rural areas have become the recipients of tailored developmental policies on several governance levels, including the European Union (Woods 2005).

Restructuring of rural areas is hence characterised by several contradictory but parallel processes of modernisation and intensification of global-local circulation, deprivation and marginalisation as well as the re-enactment of tradition and "local values" (Murdoch 2000). This happens in spatially very differentiated and distinguished forms and these spatial differentiations are always related to various forms of mobilities and immobilities, as can be seen in several examples in European countries (see e.g., Bock, Osti, and Ventura 2016, do Carmo and Santos 2012, Rau 2012, da Silva 2012).

Rural areas and their mobility characteristics

Among the EU Member States, people living in rural areas show a certain variability in terms of their relative size. On the one hand, there are countries where the percentage of inhabitants living in rural areas is around 50% (Lithuania, Denmark, Croatia, Latvia, Hungary, Slovenia, and Luxembourg); on the other, there are several with a share under 20% (Germany, Italy, Belgium, United Kingdom and the Netherlands) (Eurostat 2017).

The total number of people living in rural areas experienced a slight increase between 2010 and 2015. However, most of the member states experienced lower population growth in rural areas than in cities and the most rapid reductions in population were registered in some of the rural and sparsely populated eastern and southernmost regions of the EU (Figure 6.1).

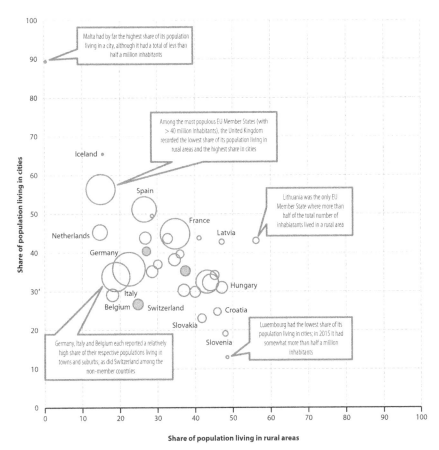

Figure 6.1 Distribution of the population by degree of urbanisation, 2015.
Source: Eurostat 2017.

In terms of employment, around 20% of the European population is employed in rural areas, underlining the importance of such regions for the European economy (European Commission 2013b). However, 81% of rural regions still have a GDP/capita below the EU average. In addition, on average, rural areas show lower income levels and lower education levels than in urban and peri-urban areas.

Also, the elderly account for a high share of the population in rural regions. This is particularly relevant as living in rural areas as an elderly person can lead to extreme forms of mobility poverty and social exclusion, especially when transport options are inadequate and when health centres and other essential services are only available in cities (Manthorpe et al. 2008, Shucksmith 2003).

The European Commission identifies a series of common problems in rural areas. These include demography (exodus of residents, ageing population), remoteness (lack of infrastructure and basic services, land abandonment), education (lack of preschools and difficulty in accessing primary and secondary schools), and labour market (low employment rates, persistent long-term unemployment, high number of seasonal workers) (European Commission 2008a).

Moreover, the rural environment is often described as deprived. Deprivation usually refers to multiple domains and ranges from income and employment to health and education, from crime and security to service barriers and social involvement.

In rural areas, mobility needs are mostly satisfied by the use of private cars. On a typical day, 64% of Europeans living in rural villages use a car as compared to 38% of residents in large towns/cities (European Commission 2015). A European survey underlines the great importance of car travel for a rural population's everyday life and reveals a strong dependence on car availability due to inadequate alternative options (European Commission 2013a).

The survey also indicates some differences in car culture. Europeans who live in urban areas are more likely than those who live in rural areas to think that additional charges for the use of specific roads at specific times would be effective in improving urban travel (45% versus 35%). The same applies to restrictions on the use of certain types of vehicle (70% versus 66%) and awareness campaigns encouraging people to limit their car use (57% versus 53%) (European Commission 2013a). Again, the survey reveals that Europeans in large towns are almost twice (51%) as likely to use public transport weekly as those in small to mid-sized towns (27%) or in rural villages (20%) (European Commission 2013a).

Of course, these figures are subject to the different levels of available and accessible public transport. For example, in rural villages, only 65% of people live less than 10 minutes away from the nearest train station or bus stop, while in large towns this percentage rises to 87% (European Commission 2013a).

Indeed, in rural areas, 20% of respondents indicated the absence/lack of public transport coverage and 18% the infrequency of public transport connections as key problems, while the percentages in urban areas are significantly lower (European Commission 2015). Insufficient availability of public transport is one of the main reasons leading to 'forced car ownership' (FCO), i.e. households being forced to own at least one car despite limited economic resources (i.e. being materially deprived) (Figure 6.2).

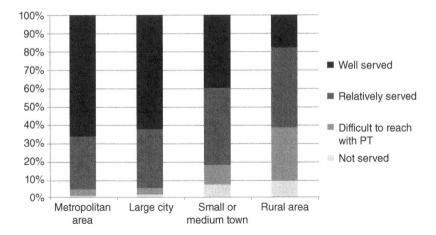

Figure 6.2 Perceived level of public transport services by level of urbanisation in the EU.

Source: European Commission 2015.

Transportation and social inclusion/exclusion

Transportation is considered as having a key role in responding to rural deprivation and rural social exclusion.

Transport policy in rural areas centres around the notion of social inclusion and is generally focused on ensuring that rural localities are inclusive of all rural citizens, including those who do not have access to private transport (Wear 2009). How to increase the mobility of all residents regardless of their socio-economic and health status, in a cost-effective way, then becomes a fundamental policy concern (Randall et al. 2018).

As regards policy advice, accessibility discourses take centre stage to inform this concern. The most critical consequence of a lack of accessibility is its contribution to social isolation for vulnerable groups. Accessibility to essential services and opportunities and access to and availability of transport options themselves are hence considered the key mechanisms for increasing the mobility of the socially and economically disadvantaged and to combat rural deprivation.

Accessibility of services is considered critical to the well-being of rural residents and the social and economic resilience of communities. Poor accessibility to services is among the factors leading to the marginalisation and peripheralisation of territories. It is argued that this may result in both a decline in economic activity and potential and low levels of socio-economic performance (low levels of well-being, quality of life, demographic ageing, economic and social stagnation) (European Network for Rural Development 2018).

The rural transport system is then understood as either inhibiting or fostering a social and economic dynamic. It is highlighted that when access to transport in rural areas is poor and of low quality, it constrains economic and social development and contributes to poverty. On the other hand, it is emphasised that when transport access is sufficient and reliable, it is a vital factor in increasing competitiveness, sustainability and the attractiveness of rural and remote areas by ensuring the accessibility of both inhabitants and potential visitors to key services such as employment, education, healthcare and leisure activities (Cook et al. 2018).

The consequence of limited transport options is particularly relevant in remote rural regions, characterised by high dispersion and numerous small villages. Such a settlement pattern is very difficult (and expensive) to be adequately served by conventional public transport services.

People living in remote and low-density rural areas inevitably have to cope with poor and infrequent public transport services that do not allow people to retain their independence and access to basic services and facilities. Limited transport links and connections make commuting almost impracticable and accelerate the depopulation of rural areas (Samek Lodovici and Torchio 2015). As mentioned, remoteness can especially impact the quality of life of groups already at risk of social exclusion and transport disadvantaged groups, such as non-car owners, the unemployed and low-income people, the elderly, women, migrants and young people.

Improving mobility and accessibility to services in rural areas are central policy responses to the challenges that rural regions are facing, especially remote regions, and are fundamental for breaking the circle of decline into which many rural areas are locked (Randall et al. 2018). This requires better transport infrastructures and mobility services that can reduce rural poverty by facilitating women, men and children to more readily access services (education, health, finance, markets), to obtain goods and income and to participate in social, political and community activities (Starkey et al. 2002).

Car dependency and forced car ownership

Large distances between services and population centres and the low population density make the provision of transport, especially considering an adequate frequency or service outside peak hours, very difficult and expensive. This means that people are strongly encouraged to almost exclusively rely on private transport and the use of their own car.

In many cases, in rural areas, using the private car is the principal and sole response to people's transportation needs. These experiences lead to the concept of 'forced car ownership' (FCO) (Mattioli 2017) used to define households who own at least one car despite limited economic resources (i.e., being materially deprived). It is assumed that these households potentially trade off motoring expenditure against expenditure in other essential areas. In households with limited resources, the enforced possession and use of a durable good can be the cause of material deprivation, economic stress and vulnerability to fuel price increases. FCO results in households cutting expenditure on other necessities and/or reducing travel activity to the bare minimum, both of which may result in social exclusion.

Beside the sustainability issues that having to exclusively rely on a private car raises, it also poses incredible challenges for the mobility of those vulnerable groups who always need to be escorted by someone else's car. Living in rural areas and being a young person, elderly or with a disability or someone who cannot independently access a car represent an extremely tough situation that can prevent access to innumerable social activities and opportunities.

In addition, it has to be considered that in rural areas around the world people are ageing. This demographic trend has a twofold effect. On the one hand, it means that there will be an increasing number of ageing drivers that will have to face the transition to becoming a non-driver. On the other, there will be an increased necessity for new kinds of services and mobility solutions to emerge. For example, fewer people will need regular public transport, whilst there will be an increasing number of people who are reliant on special transportation systems such as hospital and paratransit services (Randall et al. 2018). Alternative transport will play a key role in keeping these people engaged in mainstream society.

Policy response to the vicious circle of rural mobility poverty

While the key motivations for transport policy and mobility initiatives in urban areas are reducing congestion and pollution, the focus in rural areas is mainly on accessibility aspects, with the key question of how to increase the mobility of all residents regardless of their socio-economic and health status in a cost-effective way (Randall et al. 2018).

From a policy perspective, the rural arena is envisaged as a "vicious circle": as population density is low, the level of public transport service is generally sufficiently low to make it an unattractive alternative to the car. It is therefore not considered realistic to advance a policy goal of shifting trips from car to public transport. Hence, people in rural areas are highly dependent on cars (OECD 2009).

However, high car ownership puts pressure on existing rural public transport, prompting a diminished service, which in turn encourages even higher car ownership, creating a vicious circle of public transport decline. Also, as shown above, a transport system that is car based especially disadvantages those who are in an economically weaker position or who face a barrier due to their age or mobility impairments. These phenomena are further aggravated by processes and developments such as public budget cuts, centralisation of public services, demographic change and depopulation (Randall et al. 2018).

The success of a rural community highly depends on access to well-planned, efficient transport systems. Providing such systems is a complex policy, planning and governance goal because governments and administrations are struggling with different intersecting dynamics. One critical challenge is related to infrastructure as the 2017 report by the Council of Europe points out. The maintenance of road and rail networks is costly and ineffective due to the distances that need to be covered, the difficulties in integrating the two modes together and the very small number of users (and therefore revenues). In addition, the quality of transport infrastructure (e.g. existence of sidewalks, overall cleanliness and lighting) also tends to be inferior, or sometimes even non-existent, in rural areas. However, maintaining transport infrastructures and services is considered fundamental in order to keep those areas vital (Council of Europe 2017).

Hence, it can be concluded, from a transport policy point of view, that rural areas have always been a challenging place. Addressing rural transport issues is considered essential in order to reduce poverty and avoid social exclusion. Transportation access in rural areas is fundamental as it underpins the economic and employment development strategies of many local communities. On the other hand, providing services in these areas is very challenging. Conventional public transport is only a partial solution, while un-conventional services gain more attention in improving accessibility, as well as a seamless combination of appropriate transport infrastructure, improved transport services and affordable means of transport, both motorised and non-motorised.

Conclusions

Rural areas have always been a challenging place from a transportation point of view. Addressing their mobility issues is mandatory in order to reduce poverty and avoid social exclusion. Transportation access in rural areas is fundamental as it underpins the economic and employment development strategies of many local communities.

Rural transport enables workers to access employment and tourists to visit rural communities. More importantly, it allows local people to remain living in their town whilst accessing services or employment elsewhere (Wear 2009). Adequate infrastructure and access to transport give everyone the ability to travel and represent a fundamental resource in order to access employment and develop social relations in remote rural areas.

Therefore, a significant portion of a rural community's success depends on access to well-planned, efficient transport systems. Providing such systems is as urgent as it is breathtakingly complex. The difficulty starts with the word "rural" itself and most of the time it requires a seamless combination of appropriate transport infrastructure, improved transport services and affordable means of transport, both motorised and non-motorised.

Inclusive, participative methods involving all stakeholders are essential to determine infrastructure priorities, appropriate locations for facilities and suitable means of transport. Priorities should reflect local needs, economic development and social equity goals.

Rural transport services must be actively promoted to turn the vicious circle of insufficient transport services and inability to pay for them into a virtuous circle of better transport services that stimulate economic activity and social improvement, leading in turn to easier access and more efficient transport services. This could reduce the ongoing phenomenon of the de-population of rural areas and decrease the need for urbanisation.

Finally, it is necessary to stress the urgency to prioritise the situation of those vulnerable groups that are not capable of moving independently and autonomously. Being a young person, elderly or having a disability living in a rural area and not having the availability of suitable transportation options (aside from a private car driven by a family member) represents an extremely difficult situation. It prevents access to innumerable activities and opportunities, which consequently leads to critical social exclusion conditions. Aggravated conditions also apply to other vulnerable groups (e.g. women, those on a low income, immigrants, etc.). So, in addition to their social status' transport-related challenges, they find themselves in the situation of living in a geographically disadvantaged area from a transportation point of view.

In this sense, there are multiple inclusive mobility solutions to target vulnerable groups. However, the great majority of them are addressed to large metropolitan areas, as most of these mobility concepts require a specific infrastructure. Areas with a high population density, a mature public transportation system as well as the possibility of interconnecting different mobility options are preferred. The ultimate challenge consists in being able to successfully adapt already existing and well-established inclusive mobility solutions to serve rural needs and guarantee a brighter future for more marginalised areas as well.

References

Bell, Michael M., and Giorgio Osti. 2010. "Mobilities and ruralities: An introduction." In *Sociologia Ruralis*, 50, no.3, 199–204. https://doi.org/10.1111/j.1467-9523.2010.00518.x.

Bock, Bettina B., Giorgio Osti, and Flaminia Ventura. 2016. "Rural migration and new patterns of exclusion and integration in Europe." In *Routledge International Handbook of Rural Studies,* edited by Mark Shucksmith, David L. Brown, 101–114. Abingdon: Routledge.

Cloke, Paul. 2006. "Conceptualizing rurality." In *Handbook of Rural Studies*, edited by Paul Cloke, Terry Marsden, Patrick Mooney, 18–28. London: SAGE.

Cook, Japser, Cornie Huizenga, Rob Petts, Caroline Visser, and Alice You. 2018. "The contribution of rural transport to achieve the sustainable development goals." https://www.gov.uk/dfid-research-outputs/the-contribution-of-rural-transport-to-the-sustainable-development-goals-factsheet. Accessed 19 March 2020.

Council of Europe. 2017. "A better future for Europe's rural areas." Congress of Local and Regional Authorities. Council of Europe (33rd SESSION Report, CG33(2017)16final). https://rm.coe.int/a-better-future-for-europe-s-rural-areas-governance-committee-rapporte/168074b728. Accessed 20 June 2018.

da Silva, Vanda Aparecida. 2012. "Youth "Settled" by mobility: Ethnography of a Portuguese village." In *Translocal Ruralism*, edited by Charlotta Hedberg and Renato Miguel Do Carmo, 73–86. Dordrecht: Springer.

do Carmo, Renato Miguel, and Sofia Santos. 2012. "Between marginalisation and urbanisation: Mobilities and social change in southern Portugal." In *Translocal Ruralism*, edited by Charlotta Hedberg and Renato Miguel Do Carmo, 13–33. Dordrecht: Springer.

European Commission. 2008. "Poverty and social exclusion in rural areas." European Commission DG for Employment, Social Affairs and Equal Opportunities, Unit E2. http://ec.europa.eu/social/BlobServlet?docId=2087&langId=en. Accessed 14 June 2018.

European Commission. 2013a. "Attitudes of Europeans towards urban mobility." Special Eurobarometer 406/ Wave EB79.4. http://ec.europa.eu/commfrontoffice/publicopinion/archives/ebs/ebs_406_en.pdf. Accessed 13 June 2018.

European Commission. 2013b. "Rural development in the EU: Statistical and economic information." European Commission – Directorate-General for Agriculture and Rural Development. http://ec.europa.eu/agriculture/statistics/rural-development/2013/full-text_en.pdf. Accessed 13 June 2018.

European Commission. 2015. "EU survey on issues related to transport and mobility." Joint Research Centre. http://publications.jrc.ec.europa.eu/repository/bitstream/JRC96151/jrc96151_final%20version%202nd%20correction.pdf. Accessed 13 June 2018.

European Network for Rural Development. 2018. "Digital and social innovation in rural services." https://enrd.ec.europa.eu/sites/enrd/files/enrd_publications/publi-eafrd-brochure-07-en_2018-0.pdf. Accessed 19 March 2020.

Eurostat. 2017. "Statistics on rural areas in the EU – statistics explained." http://ec.europa.eu/eurostat/statisticsexplained/index.php/Statistics_on_rural_areas_in_the_EU#Population_distribution_by_degree_of_urbanisation. Accessed 13 June 2018.

Halfacree, Keith. 2006. "Rural space: Constructing a three-fold architecture." In *Handbook of Rural Studies*, edited by Paul Cloke, Terry Marsden, Patrick Mooney, 44–62. London: Sage.

Hedberg, Charlotta, and Renato Miguel Do Carmo. 2012. "Translocal ruralism: Mobility and connectivity in European rural spaces." In *Translocal Ruralism*, edited by Charlotta Hedberg and Renato Miguel Do Carmo, 1–9. Dordrecht: Springer.

Manthorpe, Jill, Roger Clough, Michelle Cornes, Les Bright, and Jo Moriarty. 2008. "Elderly people's perspectives on health and well-being in rural communities in England. Findings from the evaluation of the National Service Framework for Older People." In *Health & Social Care in the Community* 16, no.5, 460–468. https://doi.org/10.1111/j.1365-2524.2007.00755.x.

Mattioli, Giulio. 2017. "Forced car ownership in the UK and Germany: Socio-spatial patterns and potential economic stress impacts." In *Social Inclusion*, 5, no. 4, 147–160. http://dx.doi.org/10.17645/si.v5i4.1081.

Murdoch, Jonathan. 2000. "Networks—A new paradigm of rural development?." In *Journal of Rural studies*, 16, no.4, 407–419. https://doi.org/10.1016/S0743-0167(00)00022-X.

OECD. 2009. "Improving local transport and accessibility in rural areas through partnerships." https://www.academia.edu/9935573/Improving_local_transport_and_accessibility_in_rural_areas_through_partnerships. Accessed 19 March 2020.

Randall, Linda, Anna Berlina, Julien Grunfelder, and Arne Kempers. 2018. "Pre-study on sociocultural factors." Mamba. https://www.mambaproject.eu/wp-content/uploads/2018/08/GoA-2.6-Sociocultural-pre-study-_Nordregio.pdf. Accessed 19 March 2020.

Rau, Henrike. 2012. "The ties that bind? Spatial (im) mobilities and the transformation of rural-urban connections." In *Translocal Ruralism*, edited by Charlotta Hedberg and Renato Miguel Do Carmo, 35–53. Dordrecht: Springer.

Samek Lodovici, Manuela, and Nicoletta Torchio. 2015. "Social inclusion in EU public transport." Policy Department B: Structural and Cohesion Policies – European Parliament. Brussels. http://www.europarl.europa.eu/RegData/etudes/STUD/2015/540351/IPOL_STU(2015)540351_EN.pdf. Accessed 13 June 2018.

Shucksmith, Mark. 2003. "Social exclusion in rural areas. A review of recent research." Centre for Rural Development Research. https://www.sitesplus.co.uk/user_docs/118/File/Rural_exclusion.pdf. Accessed 19 March 2020.

Silva, Luis, and Elisabete Figueiredo. 2013. "What is shaping rural areas in Europe?" In *Shaping Rural Areas in Europe. Perceptions and Outcomes on the Present and the Future*, edited by Silva, Luis and Elisabete Figueiredo, 1–8. Dordrecht: Springer.

Smith, Darren. 2007. "The changing faces of rural populations: "(re) Fixing" the gaze' or 'eyes wide shut'?." In *Journal of Rural Studies*, 3, no.23, 275–282. https://doi.org/10.1016/j.jrurstud.2007.03.001.

Starkey, Paul, Simon Ellis, John Hine, and Anna Ternell. 2002. "Improving rural mobility. Options for developing motorized and nonmotorized transport in rural areas." World Bank Technical Paper No. 525. Washington DC: World Bank.

Wear, Andrew. 2009. "Improving local transport and accessibility in rural areas through partnerships." OECD LEED Forum on Partnerships and Local Governance. Handbook No.1.

Woods, Michael. 2005. *Rural Geography: Processes, Responses and Experiences in Rural Restructuring.* London: SAGE.

Woods, Michael. 2009. "Rural geography: Blurring boundaries and making connections." In *Progress in Human Geography*, 33, no. 6, 849–858. https://doi.org/10.1177/0309132508105001.

Woods, Michael. 2011. *Rural: Key Ideas in Geography.* New York: Routledge.

Part III
Societal roots and impacts

7 Women and gender-related aspects

Stefano Borgato, Silvia Maffii,
Patrizia Malgieri and Cosimo Chiffi

Abstract

Every day, many women struggle to access essential opportunities such as jobs, education, shops and friends. Failing to provide them with adequate transport services can eventually lead to an undesirable situation of social disadvantage and exacerbate the gender gap between men and women. This chapter explains how being a woman has a significant influence on mobility characteristics and travel behaviour. It also goes deeper into the main challenges and issues that need to be overcome to reduce inequalities of access and movement opportunities between genders. Properly addressing these challenges plays a crucial role in providing women with empowerment, access to opportunities and independence.

Introduction

When we talk about women in transportation, we are not talking about a specific group of vulnerable users. We are talking about half of the world's population. Therefore, it is fundamental to always adopt a gender-sensitive perspective when considering the particular mobility needs of people and more specifically of other vulnerable demographic groups, be they on low income, elderly, migrants, or people living in rural areas. In fact, half of these people are also women and this situation will eventually further affect the way in which they will be able to access, exploit and utilise transport services.

Recent research demonstrates that gender is one of the key factors in accounting for differences in mobility and travel behaviour, together with other important socio-demographic variables (income, household composition) as well as access to means of transport (both private and public), infrastructures and services (Peters 2013). However, it is often the least understood socio-demographic variable.

As will be highlighted in the following paragraphs, there are some significant differences between the two genders that need to be taken into account when discussing gender-related mobility. In particular, integrating gender means taking into account the different requirements and needs associated with the different genders. The concept of gender equality refers to the aim of reducing inequalities of access and opportunities between men and women (Duchène 2011).

As transport is a means to improve the well-being of people, by facilitating their access to economic and social benefits, it should be designed as equitable, affordable and responsive to all groups as possible (ADB 2013). This implies that a greater gender sensibility should be considered in mobility analysis, planning and practice (Peters 2013, CIVITAS 2014) and in designing transport policies which are often incorrectly considered as "gender neutral" (ADB 2013, Chadha and Ramprasad 2017).

However, while greater account is being taken of gender in a variety of areas, little progress has been made in this respect in the transport sector. In both the international body of literature and transport planning, the gender dimension in mobility patterns has received little attention so far, even though gender is considered a significant factor in accounting for differences in mobility behaviour. As a consequence, relevant and systematic statistical information is often not available, thus inhibiting gaining an understanding of travel practice differentiated by gender (Duchène 2011).

Most of the transport planning and policy-making all over the world are still influenced by transport planning standards, procedures and methodologies developed in industrialised countries over the course of the last century and based implicitly or explicitly on the assumption that households typically consisted of nuclear families with a traditional division of labour, i.e. a male 'breadwinner' with primary responsibility for the 'productive' tasks within the household and a female 'housewife' with primary responsibility for the 'reproductive' caretaking tasks (Sheller 2008).

In more recent decades, the evolution of household and parental models together with new developments in the labour market and new technologies with the spread of new forms of work have determined a change in the role of women in society. Women have increased their participation in the labour market, though gaps are still far from overcome. So, to some extent the travel behaviour trends of women and men are slowly converging, in particular regarding the possession of a driving licence.

However, significant differences still remain due to the different role of women and men in the in the household and the fact that women have lower employment rates, the majority of part-time roles and low-wage positions. As a consequence, women have far more complex activity schedules: household chores and childcare require a higher degree of synchronisation, planning and coordination with multiple external factors. In other words, these mobility needs require a greater effort in order to be addressed.

Clarifying female (and male) mobility trends and patterns should help understand the persistence of gender roles, which is necessary for policymakers to better target directives aimed at addressing women's mobility challenges (McGuckin and Nakamoto 2004).

Socio-demographic characteristics

As outlined in the introduction, women represent a very large vulnerable group as they account for 51% of the total European population. Therefore, when we discuss gender transportation challenges, we are talking about an issue that affects more than half of Europe's citizens. Indeed, there are some significant differences in terms of socio-demographic characteristics between males and females that is worth mentioning in order to better comprehend women's transportation behaviour and mobility challenges.

Notably, women leave their parental home earlier than men (around two years earlier) and get married earlier (between three and two years earlier). In addition, women live longer than men, with an average 5.4 years difference in 2016 (Eurostat 2018a).

As a result of a longer life expectancy, there are more women than men in the EU, with 105 women per 100 men in 2017 (Eurostat 2018a). The biggest differences can be seen in Latvia (18% more), Lithuania (17% more) and Estonia (13% more), while Luxembourg, Malta and Sweden have slightly more men than women (Eurostat 2018a). Looking at young people aged up to 18, the opposite pattern applies with 5% more young men than young women of this age. On the other hand, among the older group aged 65 and over, there are 33% more women (Eurostat 2018a).

Other differences can be observed considering household compositions. In 2017 in the EU, around 8% of women aged 25–49 lived alone with children, compared with 1.1% of men of the same age. For singles without children in this age group, the percentage was 9.6% for women and 16.3% for men. Another group with large differences between men and women is for singles aged 65 and over: the percentage of elderly women living alone (40.4%) was twice the percentage for men (19.9%) (Eurostat 2018a).

When looking at the level of education completed, there are hardly any differences between women and men in the EU at lower education level. However, different patterns can be seen at the higher levels. Specifically, there is a majority of women (33% versus 30%) who have completed the tertiary level of education in almost all Member States (Eurostat 2018a).

On average, the employment rate of men is higher than that of women (73% compared with 62% in the EU in 2017). However, this difference increases with the number of children. The employment rate for women without children is 66%, while it is 74% for men. For women with one child, the rates increase and are 71% for women and 86% for men. For women with two children, the rate remains almost the same at 72%, while the one for men increases to 90%. Finally, for those with three or more children, the employment rate drops to 57% for women, compared to 85% for men (Eurostat 2018a).

The principal reasons behind this employment gap may be (Eurostat 2018b):

- Labour market issues, including employers preferring to hire young men over young women; young women facing assimilation difficulties when returning to work after childbirth; young women being more likely to have low-paid jobs or precarious employment.
- The way the family work and childcare is currently divided between parents, with a great majority of females undertaking such activities.
- Social conventions, which tend to place a higher importance on women's role within the family.
- Education and careers advice, which often reinforce gender segregation and direct women into a relatively narrow range of occupations.

In addition, an important aspect of the reconciliation between work and family life is part-time work. However, this is not equally divided between women and men. 32% of women in employment work part-time, compared with 9% of men. The highest share of women working part-time is in the Netherlands (74%), while the lowest is observed in Bulgaria (2%) (Eurostat 2018a). Also, the female/male unemployment rates show some differences between the Member States, but the EU average is very close (7.9% for women versus 7.4 for men) (Eurostat 2018a).

In terms of earnings, women earn 16.2% less than men when comparing their average gross hourly earnings, giving an overall picture of gender inequalities in terms of hourly pay. That said, this pay gap is linked to a number of cultural, legal, social and economic factors which go far beyond the single issue of equal pay for equal work (Eurostat 2018a).

Finally, for all Member States, there is a much larger share of women undertaking child care, housework and cooking than men. 92% of women aged 25 to 49 (with children under 18) take care of their children on a daily basis, compared with 68% of men. The largest differences are observed in Greece (95% of women versus 53% of men) and Malta (93% and 56%, respectively), while the smallest are in Sweden (96% of women and 90% of men) and Slovenia (88% and 82%, respectively) (Eurostat 2018a).

Gender differences in travel patterns

Gender is an important factor in accounting for notable differences in mobility and travel behaviour. Since women and men experience transport differently, as they use different modes for different purposes and in different ways, they also have different preferences and constraints.

Recognition of the existing links between gender and mobility has only recently begun to emerge in literature. One of the reasons is the lack of gender-differentiated statistics that makes it difficult to understand gender differences in relation to reasons for making journeys, journey frequencies, distance travelled and mobility-related problems in accessing services and employment.

That said, it is possible to affirm that the major differences in the basic mobility needs of women and men are grounded in the different social roles they play and in the gender-based division of labour within the family and the community (CIVITAS 2014).

Statistics show that women spend more than two thirds of their time at home. They also usually have less free time than men, being engaged in childcare, domestic work and caring for elderly, sick or disabled relatives. Therefore, women are more likely to work part-time, to choose jobs that are nearer or better connected to home (even if low-paid) or to decide not to work at all (Eurostat 2018a).

A review of the literature showed significant gender differences in terms of frequency, time, mode and purpose of travel (Hasson and Polevoy 2011, Schwanen 2011, Hodgson 2012, Samek Lodovici et al. 2012, CIVITAS 2014, Department for Transport 2014, Tilley and Houston 2016).

In particular, these studies have indicated that women with respect to men tend to:

- Have shorter commutes;
- Have a shorter distribution of travel during the day and less concentrated during peak hours;
- Transform chain trips into complex journeys to conduct household-serving trips more often; and
- Make less use of the car and more of public transportation.

On average, women commute shorter than men in both distance and time due to lower incomes, available modes of transport, occupation status, location choice, socio-economic factors, geographical structure and infrastructure availability.

In general, women face greater time-space constraints in commuting. In 2015, in the EU the share of male and female outbound commuters among all employed persons was systematically higher for men compared to women in each of the member states for which data were available (Eurostat 2018b).

In addition, while on average women tend to travel less for work (and these differences tend to increase together with the gender disparities in labour market participation), they travel more frequently for the purposes of shopping, escorting family members and household management (CIVITAS 2014) (Figure 7.1).

In terms of mode choice, there is a gender difference in relation to the utilisation of different transport modes (European Commission 2013). In fact, a higher proportion of men travel by car and motorcycle, while women walk, bike and use public transport more than men.

Men are more likely than women to use a car daily (57% versus 42% at EU level) (European Commission 2013). On the other hand, women tend to

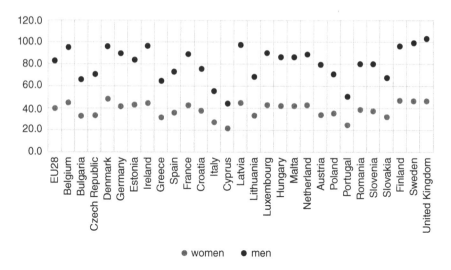

Figure 7.1 Average duration of commuting time one-way between work and home by gender (15–64), 2015.
Source: Eurofound 2017.

travel in cars more frequently as passengers rather than as drivers (CIVI-TAS 2014). Also, women have less access to private cars and driving licences than men for reasons that may be attributed to economic inequality and gender stereotypes. In single car households, the car is normally used by the male partner.

That said, the number of women drivers has been growing between 2000 and 2010 (a 3.5% increase) (SARTRE 2012) and women's access to a car is increasing almost universally across all age groups (Stokes 2012). The converging patterns are also shown with regard to the possession of a driving licence (Duchène 2011).

While women have lower rates of motorisation than men, they complete a larger share of trips by walking or using the bicycle (Heesch et al. 2012,). Also, women tend to use public transport more than men (Allen 2018).

Considering shared mobility, several surveys underlined that the regular user of car-sharing and bike-sharing is more likely to be male rather than female (Pickup et al. 2015, Chatterjee et al. 2018).

In terms of older people, researchers indicate that older women's travel patterns and transport choices have been changing as well, reducing differences with older men's patterns (Department for Transport 2009, Su and Bell 2012).

Finally, compared to men, women tend to be more environment-focused as well as hold more positive views of speed limits and congestion fees and initiatives geared towards the promotion of a more sustainable transport system (Basarić et al. 2016).

Challenges for women in transportation

Even if gender mobility patterns have been changing in recent years, reflecting the evolution of gender differences in socioeconomic and demographic conditions (CIVITAS 2014), the analysis of the socio-economic background together with projections and trends confirm that, though narrowing, the gap between genders is still evident and has effects on mobility patterns.

As highlighted in the previous section, differences in participation in the labour market, the employment gap, caring and family duties and education level influence women's mobility patterns and pose a series of challenges in order to satisfy them. Transport is still not gender-neutral and gender needs have to be properly addressed by experts and policymakers (Allen 2018).

It is fundamental to properly address these challenges as transportation access plays a crucial role in empowerment, access to opportunities, social life and independence (Samek Lodovici and Torchio 2015, World Economic Forum 2017). Research shows that poor mobility and access to transport can prevent women from entering the labour market or lead women to choose less profitable jobs because they are closer to home or easier to travel to, even in the case of self-employment (Hanson 2003).

In order to guarantee fair access to transport for both men and women and reduce the disadvantages for the latter, a series of considerations should be taken into account in order to guarantee the quality, safety and comfort measures required by women.

First of all, we should consider that women have generally less access to resources and are economically-disadvantaged regarding the control of the household's finances. Therefore, affordability of transport is especially relevant for them. Constrained access to transport can also exacerbate gendered poverty. Women are likely to take inexpensive and therefore slower modes of transport. Therefore, it is essential to improve the overall conditions for travelling to allow women to meet their needs and aspirations as well (Allen 2018).

As mentioned, women have different travel patterns and use public transport and walking more than men. They make more trip chains and depend more on off-peak and off-branch travel. This needs to be understood and taken into account by removing the barrier that might prevent women from accessing public transport (e.g. many women travel with children and strollers) and by offering more flexible services (Allen 2018).

In this respect, the traditional public transit offer, with schedules concentrated on peak hours to primarily satisfy journey-to-work trips and on high demand links, does not generally fit women's needs and basically disregards the needs of part-time/shift working or non-working people.

In addition, it is fundamental to take into account the fact that women's modal choice depends not only on the conventional parameters (times, costs and comfort) but also on the conditions and safety of the journey. Women are more affected by safety and security issues as they are more likely to encounter violence and harassment when they are using public space, particularly on public transport. This reduces the freedom of movement of women and girls, and their ability to attend school or work and participate fully in public life.

Failure to take account of women's safety sometimes prompts the latter to prefer private car use to public transport or limit their presence on public transport to certain hours and certain routes that are perceived as safe. A bus user survey conducted in London in 2008 showed that women are far less likely to take a night bus (35%) than they are to take a day bus (54%) (Transport for London 2014). The failure to take into consideration women's safety can encourage them to choose their private car instead, if available, or to forego the trip, if not.

Finally, it has to be taken into account that, considering the current ageing trends – women account for the majority of elderly people – and the higher life expectancy at birth for girls, there will be an increasing number of old women living alone, with significant mobility problems and difficulties in accessing services.

Conclusions

As women represent half of the world's population and constantly face multiple challenges when it comes to mobility (whether related to inclusion, safety or accessibility), it is fundamental to properly address such difficulties, especially when they are combined with the ones typical of other vulnerable groups (e.g. women living in rural areas, low-income women, female migrants).

Lower employment rates, part-time roles, family care duties, low-wage positions and safety issues are the main factors which determine sensible differences between genders in the labour market, in social life and consequently in transport behaviour. The emerging picture is one where women travel differently than men in relation to transport mode used, distance travelled, the daily number of trips and their pattern and travel purposes (CIVITAS 2014).

In order to reduce women's burden and address their diverse requirements, further study is mandatory to better understand women's mobility. Collecting gender-disaggregated data to understand female travel patterns and conducting gender impact assessments should be the starting point that will eventually encourage planners, policymakers and service managers to embrace gender-responsive policies and develop more inclusive (friendly street network, barrier-free public transport) and safe (public areas with visibility, lighting at transit stations, trained transit staff to deal with harassment) mobility environments and services for women.

Mobility and access have to be recognised as essential conditions equally for both women and men to be able to exercise many of their rights, tasks and activities, including access to work, education, social activities and other essential services. It is now time to think about how to really make mobility systems gender-responsive, sustainable and affordable, giving women real choices and access to opportunities that will avoid the social exclusion of women in vulnerable situations and will achieve equity between women and men.

References

ADB. 2013. "Gender tool kit: Transport maximizing the benefits of improved mobility for all." Asian Development Bank, Mandaluyong City, Philippines. https://www.adb.org/sites/default/files/institutional-document/33901/files/gender-tool-kit-transport.pdf. Accessed 19 March 2020.

Allen, Heather. 2018. "Approaches for Gender Responsive Urban Mobility." Sustainable Urban Transport Project. Bonn and Eschborn: Deutsche Gesellschaft für Internationale Zusammenarbeit (GIZ) GmbH.

Basarić, Valentina, Jelena Mitrović, Vuk Bogdanović, and Nenad Saulic. 2016. "Gender and age differences in the travel behavior – a Novi Sad case study." In *Transportation Research Procedia* 14: 4324–4333. https://doi.org/10.1016/j.trpro.2016.05.354.

Chadha, Jyot, and Vishal Ramprasad. 2017. "Why it is key to include gender equality in transport design." In *The City Fix*. https://thecityfix.com/blog/why-it-is-key-to-include-gender-equality-in-transport-design-jyot-chadha-vishal-ramprasad/. Accessed 8 March 2017.

Chatterjee, Kiron, Phil Goodwin, Tim Schwanen, Ben Clark, Juliet Jain, Steve Melia, et al. 2018. "Young people's travel – what's changed and why? Review and analysis." The Centre for Transport & Society. UWE Bristol, UK. www.gov.uk/government/publications/young-peoples-travel-whats-changed-and-why. Accessed 13 June 2018.

CIVITAS. 2014. "Smart choices for cities – Gender equality and mobility: mind the gap!," CIVITAS WIKI Policy Analyses series. http://www.ricerchetrasporti.it/test/wp-content/uploads/downloads/2014/10/CIVITAS_Second-Policy-Note_Gender-equality-and-mobility-mind-the-gap.pdf. Accessed 19 March 2020.

Department for Transport. 2009. UK, National household travel survey. Department for Transport. https://www.gov.uk/government/collections/national-travel-survey-statistics#publications. Accessed 20 June 2018.

Department for Transport. 2014. National travel survey trip chaining: 2002–2014 (UK Government Factsheet 2014). https://www.gov.uk/government/uploads/system/uploads/attachment_data/file/509447/nts-trip-chaining.pdf. Accessed 14 June 2018.

Duchène, Chantal. 2011. "Gender and transport." OECD/ITF Joint Transport Research Centre Discussion Papers – Discussion Paper No. 2011-11. http://www.oecd-ilibrary.org/docserver/download/5kg9mq47w59wen.pdf?expires=15030

70935&id=id&accname=guest&checksum=1841AC9F8A1703AF30ADE9A6F 22EF00C. Accessed 19 March 2020.

Eurofound. 2017. 6th European working conditions survey. 2017 update. http:// rhepair.fr/wp-content/uploads/2017/12/2017-Update-6th-European-Working-Conditions-Survey-Eurofound.pdf. Accessed 20 June 2018.

European Commission. 2013. "Attitudes of Europeans towards urban mobility." Special Eurobarometer 406/ Wave EB79.4. http://ec.europa.eu/commfrontoffice/ publicopinion/archives/ebs/ebs_406_en.pdf. Accessed 13 June 2018.

Eurostat. 2018a. "The life of women and men in Europe." https://ec.europa.eu/ eurostat/cache/infographs/womenmen/images/pdf/WomenMenEurope-Digital Publication-2018_en.pdf?lang=en. Accessed 19 March 2020.

Eurostat. 2018b. Database extraction. http://eige.europa.eu/gender-statistics/dgs. Accessed 20 June 2018.

Hanson, Susan. 2003. "Geographical and feminist perspectives on entrepreneurship." In *Geographische Zeitschrift* 91, no. 1: 1–23. https://doi.org/10.2307/27818963.

Hasson, Yael, and Marianna Polevoy. 2011. "Gender equality initiatives in transportation policy, a review of the literature." Women's Budget Forum, Heinrich Boell Stiftung, European Union, Hadassah Foundation. https://il.boell.org/sites/ default/files/gender_and_transportation_-_english_1.pdf. Accessed 14 June 2018.

Heesch, Kristiann C., Shannon Sahlqvist, and Jan Garrard. 2012. "Gender differences in recreational and transport cycling: A cross-sectional mixed-methods comparison of cycling patterns, motivators, and constraints." In *International Journal of Behavioral Nutrition and Physical Activity* 9: 106. https://doi.org/10.1186/ 1479-5868-9-106.

Hodgson, Frances. 2012. "Everyday connectivity: equity, technologies, competencies and walking". In *Journal of Transport Geography* 21: 17–23. https://doi.org/10. 1016/j.jtrangeo.2011.11.001

McGuckin, Nancy, and Yukiko Nakamoto. 2004. "Differences in trip chaining by men and women." In *Conference Proceedings 35: Research on Women's Issues in Transportation*, Volume 2: Technical papers, 49–45. http://onlinepubs.trb.org/ onlinepubs/conf/CP35v2.pdf. Accessed 27 March 2020.

Peters, Deike. 2013. "Gender and sustainable urban mobility. Official thematic study for the 2013 UN Habitat Global Report on Human Settlements." https:// unhabitat.org/wp-content/uploads/2013/06/GRHS.2013.Thematic.Gender.pdf. Accessed 19 March 2020.

Pickup, Laurie, Oriol Biosca, Laurent Franckx, Herman Konings, Inge Mayeres, Pnina Plaut et al. 2015. "MIND-SETS: A new vision on European mobility." Deliverable 2.1A of the MIND-SETS project. European Commission Directorate General for Research. http://www.mind-sets.eu/wordpress/wp-content/ uploads/2015/11/D2.1.a_final.pdf. Accessed 13 June 2018.

Samek Lodovici, Manuela, and Nicoletta Torchio. 2015. "Social inclusion in EU public transport." Policy Department B: Structural and Cohesion Policies – European Parliament. Brussels. http://www.europarl.europa.eu/RegData/etudes/ STUD/2015/540351/IPOL_STU(2015)540351_EN.pdf. Accessed 13 June 2018.

SARTRE. 2012. "European road users' Risk perception and mobility." The SARTRE 4 Survey, Report to the European Commission. Champs-sur-Marne: IFSTTAR.

Schwanen, Tim. 2011. "Car use and gender. The case of dual-earner families in Utrecht, The Netherlands." In *Auto Motives. Understanding Car Use Behaviours,* edited by Karen Lucas, Evelyn Blumenberg, Rachel Weinberger, 151–171. Oxford: Emerald Group Publishing Limited.

Sheller, Mimi. 2008. "Gendered mobilities. Epilogue." In *Gendered Mobilities*, edited by Tanu P. Uteng and Tim Cresswell, 257–265. Aldershot: Ashgate.

Stokes, Gordon. 2012. "Has car use per person peaked? Age, gender and car use." Presentation to Transport Statistics User Group. London. April 2012. http://www.gordonstokes.co.uk/transport/peak_car_2012.pdf. Accessed 19 March 2020.

Su, Fengming, and Michael G. H. Bell. 2012. "Travel differences by gender for older people in London." In *Research in Transportation Economics* 34, no. 1, 35–38. https://doi.org/10.1016/j.retrec.2011.12.011.

Tilley, Sara, and Donald Houston. 2016. "The gender turnaround. Young women now travelling more than young men." In *Journal of Transport Geography* 54, 349–358. https://doi.org/10.1016/j.jtrangeo.2016.06.022.

Transport for London. 2014. "Understanding the travel needs of London's diverse communities. A summary of existing research." http://content.tfl.gov.uk/understanding-the-travel-needs-of-london-diverse-communities.pdf. Accessed 13 June 2018.

World Economic Forum. 2017. "The global gender gap report 2017." World Economic Forum. Geneva. www.weforum.org. Accessed 14 June 2018.

8 People on low income and unemployed persons

Stefano Borgato, Silvia Maffii and Simone Bosetti

Abstract

For obvious reasons, low-income people constantly face serious difficulties in accessing transport. On one hand, affordability represents a financial burden in purchasing transportation services. On the other, due to the areas in which they tend to live or irregular mobility patterns, low-income people are often forced to purchase a private vehicle, despite having very limited economic resources. This chapter analyses the existing correlation between material poverty and mobility poverty. It also explores how people's mobility behaviour is influenced by their economic status, which is often connected to the range of suitable transport alternatives available. More transport options for low-income people mean greater access to opportunities, higher chances of finding better jobs and ultimately not remaining further excluded from society.

Introduction

Poverty, with the meaning of material deprivation and directly linked to people's disposable income and level of employment, is not only the most widespread form of poverty, but also a multidimensional issue that can be analysed from different perspectives using different indicators. In particular, this chapter establishes the relationship which exists between material poverty, unemployment and poverty in transport. This is achieved while attempting to understand how mobility behaviour and opportunities change due to people's income and social position.

Being able to move and having access to transport entail certain costs. In modern and industrialised economies, the proportion of these costs of a household's budget can be considerable. Certainly, these costs vary according to multiple factors, including employment (commuters tend to spend more on transport than people who don't commute), income (lower-income households tend to spend less in total but more as a proportion of their income than higher income households), vehicle ownership (vehicle owning households tend to spend a greater proportion of their income than zero-vehicle households), geography (suburban and rural households spend more than urban households) and the quality of local transport options (residents of neighbourhoods with better mobility options tend to spend less than those in automobile-dependent communities) (Litman 2017).

At the same time, it is widely believed that mobility is a key element in terms of economic and social opportunities. The link between mobility and transport disadvantages is profound and has an important impact in shaping quality of life for both individuals and communities. Transport disadvantage and lower levels of access to modes of transport are also linked to social exclusion and poor access to goods, services and jobs. This link can contribute to social exclusion making it difficult for people to fully participate in society (Titheridge et al. 2014; Lucas et al. 2016).

In particular, low-income people constantly face a "transport affordability" problem, i.e. a financial burden when purchasing transportation services, particularly those required to access basic goods and activities such as school, work, healthcare, shopping and social activities. The problem also occurs when a household is forced to consume more travel costs than it can reasonably afford, especially costs relating to car ownership and usage. In this sense, low-income people can often be categorised as 'obliged' to possess a private car. This is either because they live in areas that are generally less well served by transport services or because they tend to have less foreseeable mobility patterns (multiple part-time and irregular jobs, seeking low cost goods and services even if less convenient). This leads to the important concept of 'forced car ownership' (FCO), used to define households who own at least one car despite very limited economic resources (i.e. being materially deprived) (Mattioli 2017).

Material poverty in the European Union

Tackling material poverty reduction is a key policy component of the Europe 2020 Strategy within the EU's agenda for growth and jobs for the current decade (European Commission 2010). In this sense, the EU's progress in reducing poverty is monitored by the headline indicator "people at risk of poverty or social exclusion" (an AROPE indicator).

This indicator takes into account three different dimensions of poverty:

- People at risk of poverty after social transfers;
- Severely materially deprived people; and
- People living in households with very low work intensity.

According to Eurostat's EU statistics on income and living conditions (EU-SILC) (Eurostat 2016), in 2015, almost 119 million people (23.7% of the total EU population) were at risk of poverty or social exclusion.

The development of the risk of poverty or social exclusion in the EU over the past decade has been marked by two turning points: in 2009, when the number of people at risk started to rise because of the delayed social effects of the economic crisis and, in 2012, when this upward trend reversed. By 2015, the number of people at risk had almost fallen to the 2008 level, reaching 118.8 million people (Eurostat 2017).

In addition, almost 39 million people, or nearly one third (32.5%) of all people at risk of poverty or social exclusion, were affected by more than one dimension of poverty over the same period. Another 9.2 million people, or 1 in 12 of those at risk of poverty or social exclusion (7.7%), were affected by all 3 forms (Eurostat 2017).

That being said, these average figures, calculated as a weighted average of national results, mask considerable variations between EU Member States. In 2015, more than a third of the population was at risk of poverty or social exclusion in three EU Member States: Bulgaria (41.3%), Romania (37.3%) and Greece (35.7%). At the other end of the scale, the lowest shares of persons at risk of poverty or social exclusion were recorded in Finland (16.8%), the Netherlands (16.4%), Sweden (16.0%) and the Czech Republic (14.0%).

Three southern European countries – Greece, Cyprus and Spain – experienced the most substantial increases in their share of people at risk of poverty or social exclusion from 2008 to 2015, ranging from five to eight percentage points (Figures 8.1 and 8.2).

Monetary poverty is the most widespread form of poverty, affecting, in 2015, 17.3% of the EU population, who earned less than 60% of their respective national median equivalised disposable income, the so-called poverty

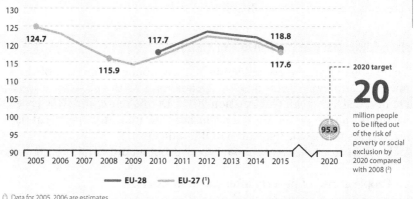

(¹) Data for 2005, 2006 are estimates
(²) The overall EU target (referring to the EU-27 — the 27 EU countries before the accession of Croatia) is to lift at least 20 million people out of the risk of poverty or social exclusion by 2020. Due to the structure of the survey on which most of the key social data is based (EU Statistics on Income and Living Conditions), a large part of the main social indicators available in 2010, when the Europe 2020 strategy was adopted, referred to 2008 as the most recent year of data available. For this reason progress towards the Europe 2020 strategy's poverty target is monitored using 2008 as the baseline year.
Source: Eurostat (online data code: t2020_50)

Figure 8.1 People at risk of poverty or social exclusion, EU-27 and EU-28, 2005–2015 (million people).
Source: Eurostat 2017.

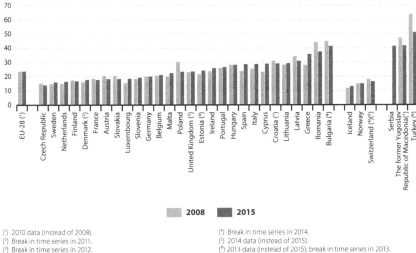

Figure 8.2 People at risk of poverty or social exclusion, by country, in 2008 and 2015 (% of population).

Source: Eurostat 2017.

threshold. This represents a slight increase compared with 2008, when 16.5% fell below this threshold.

The rate for the EU-28, calculated as a weighted average of national results, conceals considerable variations across the EU Member States. In eight Member States, namely Romania (25.4%), Latvia (22.5%), Lithuania (22.2%), Spain (22.1%), Bulgaria (22.0%), Estonia (21.6%), Greece (21.4%) and Croatia (20.0%), one fifth or more of the population was identified as being at risk of poverty. Among the Member States, the lowest proportions of persons at risk of poverty were observed in the Czech Republic (9.7%) and the Netherlands (11.6%).

Most countries also experienced growth in the number of people below the monetary poverty line, regardless of whether they already had low or high levels of monetary poverty. Increases were most pronounced in Hungary, Sweden and Spain, with rises of between 2.3 and 2.5 percentage points. Croatia, Finland, Austria, the United Kingdom and Latvia were the exception, with monetary poverty in these countries decreasing by 0.6–3.4 percentage points between 2008 (Croatia: 2010) and 2015.

The differences in poverty rates are more pronounced when the population is classified according to activity status. The unemployed are a particularly vulnerable group: almost half (47.5%) of all unemployed persons in the EU-28 were at risk of poverty in 2015, with by far the highest rate in Germany (69.1%), while seven other EU Member States (the three Baltic States,

Bulgaria, Hungary, Romania and Malta) reported that at least half of those unemployed were at risk of poverty in 2015.

Even among the employed, there is a high risk of social exclusion for workers with low-quality jobs (precarious, low-paid, part-time jobs) resulting in low or no income and a high risk of falling into poverty and material deprivation.

Around one in eight (13.2%) retired persons in the EU-28 were at risk of poverty in 2015; rates that were at least twice as high as the EU-28 average were recorded in Lithuania (27.6%), Bulgaria (30.0%), Latvia (36.7%) and Estonia (40.1%).

Those in employment were far less likely to be at risk of poverty (an average of 9.5% across the whole of the EU-28 in 2015). There was a relatively high proportion of employed persons at risk of poverty in Romania (18.8%) and to a lesser extent in Greece (13.4%) and Spain (13.1%), while Luxembourg, Italy, Poland and Portugal each reported that in excess of 1 in 10 members of their respective workforces were at risk of poverty in 2015.

Paid employment is crucial for ensuring sufficient living standards and it contributes to economic performance, quality of life and social inclusion, making it one of the cornerstones of socioeconomic development and well-being.

In 2016, 71.1% of the EU population aged 20–64 were employed. This is by far the highest share that has been observed since 2002. However, it is still 3.9 percentage points behind the EU 2020 employment target of 75%. In 2016, 6.5% of the population were unemployed; the remaining 22.5% were inactive, meaning they were not (actively) looking for work (Eurostat 2018a).

Employment rates across the EU tend to show a north-south divide at a country as well as regional level. Some of the best performing countries such as Germany, Sweden and the United Kingdom also record high regional employment rates.

In Scandinavian and western European countries, employment rates tend to be higher in rural areas. Whereas in most Baltic, southern, central or eastern Member States, cities exhibit higher employment rates.

Considerably lower employment rates are observed for women than men. The gender employment gaps are widest for women in age groups associated with having caring responsibilities for children, dependent family members or grandchildren.

Among the Member States, the lowest unemployment rates in February 2018 were recorded in the Czech Republic (2.4%), Germany and Malta (both 3.5%) as well as Hungary (3.7% in January 2018). The highest unemployment rates were observed in Greece (20.8% in December 2017) and Spain (16.1%) (Figure 8.3).

Youth unemployment rates are generally much higher, even double or more than double, than unemployment rates for all ages.

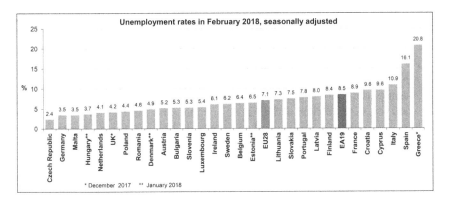

Figure 8.3 Unemployment rates, seasonally adjusted, February 2018 (%).
Source: Eurostat 2018b.

With an unemployment rate of 18.7% in 2016, young people aged 15–29 were clearly at a disadvantage compared with the overall population.

Over the past few years, increases in part-time work and fixed-term contracts have been observed. Young people have been the most affected, with 16% of 15–24-year-olds involuntarily employed on time-limited contracts and 8.4% involuntarily in part-time work in 2016.

The proportion of people at risk of monetary poverty is also closely linked to income inequality. Data on economic inequality become particularly important for estimating relative poverty because the distribution of economic resources may have a direct bearing on the extent and depth of poverty.

In 2015, there were wide inequalities in the distribution of income: a population-weighted average of national figures for each of the individual EU Member States shows that the top 20% of the population (with the highest equivalised disposable income) received 5.2 times as much income as the bottom 20% (with the lowest equivalised disposable income).

This ratio varied considerably across the Member States from 3.5 in Slovakia and the Czech Republic to 6.0 or more in Portugal, Estonia, Latvia, Greece, Spain, Bulgaria and Lithuania, peaking at 8.3 in Romania.

Income-related differences in transport and mobility

Many researchers, from both North America and Europe, have attempted to understand, from multiple perspectives, the relationship between poverty, low income, unemployment and access to transport (affordability, private car dependence). Not surprisingly, their studies have found that low-income people have less access to private modes of transport and are more likely to use public transport. They also tend to travel shorter distances and are more sensitive to public transport fares.

One of the principal results is that low-income and unemployed people are particularly reliant on local public transport services, since in many cases they cannot afford a private car or other means of transport. Unemployed people are the category that is most likely to use public transport: 23% compared to an average 19% (European Commission 2014).

A survey showed that while 52% of people who almost never have difficulties in paying their bills use the car on a daily basis, this percentage goes down to 37% for those who report difficulties in paying their bills (European Commission 2013).

In terms of car availability, in general the lower the income, the lower the availability of cars (European Commission 2015). Individuals belonging to high-income households do have a higher number of cars available in comparison to those living in low-income households.

On the other hand, those on low income and unemployed living in remote, peripheral or deprived areas often have to rely on private vehicles to access essential services, posing a substantial financial burden on these households. This leads to the already-introduced phenomenon of 'forced car ownership', where households have to own at least one car despite limited economic resources (i.e. being materially deprived). It is assumed that these households potentially trade off motoring expenditure against expenditure in other essential areas. In households with limited resources, the enforced possession and use of a durable good can be the cause of material deprivation, economic stress and vulnerability to fuel price increases. FCO results in households cutting expenditure on other necessities and/or reducing travel activity to the bare minimum, both of which may result in social exclusion. In 2012, 6.7% of UK households and 5.1% in Germany were classified as FCO (Mattioli 2017).

When it comes to travel intensity, the average number of trips made per weekday rises with increasing household income. For Londoners with an annual household income of less than £20,000, the average number of trips per weekday is 2.40 and for Londoners with a household income of below £5,000, the average number of trips made per weekday is 2.21, compared to 2.68 for all Londoners (Transport for London 2014).

Mobility challenges for low-income persons

As seen in the section above, income – either individual or at household level – is highly influential on travel behaviour. The travel behaviour (and transport conditions) of lower income groups has very specific patterns highly differentiated from their higher income counterparts in almost every country of the world (Lucas et al. 2016). A clear recognition of these differences is extremely important for the planning and delivery of economically, environmentally and socially sustainable transport systems.

First of all, low-income individuals tend to be less mobile, as they suffer from a lack of both private and public transport services in terms of the number of options and quality of the services available to them. They limit themselves to compulsory trips while other trips, most of the time of a social nature (e.g. visiting friends, relatives), are reduced, if not eliminated (Moore et al. 2013). Such behaviour can be observed worldwide in many countries (Lucas et al. 2016).

At the same time, in urban areas poor people most often live in peripheral locations at the edges of cities with a low amenity value, where there are few local employment opportunities and an absence of local services and basic facilities. Together with limited access to transport options they produce a 'poverty trap', curtailing access to jobs, education, health facilities and social networks (Lucas et al. 2016).

Furthermore, due to budgetary reasons, vulnerable segments often suffer from a lack of access to private and public transport services in terms of both options and quality of service (Barter 1999). They are then forced to rely on walking or cycling, which inevitably reduces the amount and scope of opportunities they can reach. The quality of transport infrastructure (e.g. existence of sidewalks, overall cleanliness or illumination) also tends to be inferior in deprived regions. Hence, low-income groups are often confronted with the need to walk and cycle in unsafe conditions for longer periods and routes. Therefore, there is a higher risk of road casualties and exposure to pollutants with a direct impact on their quality of life and well-being (Titheridge et al. 2014).

It is also important to consider that transport needs and habits depend on a neighbourhood's level of accessibility and social expectation, as it can be experienced particularly in North America. In more accessible neighbourhoods, it may be relatively easy to live without driving a car. Non-motorists do not face social isolation, transport's financial costs tend to be relatively low and driving is considered a convenience. On the contrary, in a more car-dependent neighbourhood, driving is unavoidable, due to scattered destinations, poor alternative travel options and because alternative modes (walking, cycling and public transport) have a bad reputation. In such localities, non-motorists tend to experience social isolation and transport's financial costs are higher. Consequently, lower-income households are relatively poorer and experience more hardship and loss of social esteem (Litman 2003).

Finally, the poorest sections of society do not benefit equally from new or improved mobility services. This may either be because they do not have access to private automobiles or because they cannot afford public transport services. The result is that the poorest population groups may become even further marginalised and sink into poverty (Lucas et al. 2016).

Conclusions

This chapter has analysed – from different perspectives – the existing correlation between material poverty and mobility poverty and described how low-income people's social status affects their mobility behaviour, travel choices and ultimately their ability to access opportunities, including school, work, healthcare, shopping, leisure and social activities.

The availability and affordability of means of transport are essential to access these opportunities in order to avoid exclusion from society. However, the necessity of being mobile can represent a serious burden for low-income and unemployed people. As seen in the previous section, it is indisputable that this category of individuals faces several mobility challenges and can be considered transport-disadvantaged in multiple ways.

For this reason, there is an urgent need for transport policymakers to carefully consider targeted interventions aiming at improving accessibility for those on low income and unemployed so that they can have more opportunities for social interactions, higher chances of finding better jobs and ultimately not remain excluded from society.

A first key issue that should be addressed is around the role that the car plays for low-income individuals. On one hand, people who cannot afford a car face significantly higher risks to social exclusion. This is especially true for low-income people living in rural areas where public transport availability is insufficient and distances to services and opportunities are larger than in urban areas. On the other hand, as car ownership is almost unavoidable in certain areas, its forced possession (and usage) poses a significantly high cost burden on already materially deprived individuals. The money spent on mobility is then missing in other essential areas of life, thus threatening even further their poverty status and social exclusion risk.

A second key element to be addressed is related to the role of public transportation and how fundamental this resource is to low-income population groups, especially those living in urban areas. Therefore, activities to promote or enhance such services could certainly have a noteworthy impact on the possibility of low-income individuals to reach opportunities and overcome isolation.

Finally, as experiences of transport disadvantages are often associated with material deprivation, it must be assumed that a large part of those at risk of poverty are also at risk of mobility poverty. However, it is relevant to bear in mind that it is not only material poverty that affects mobility poverty. In fact, the risk of social exclusion is particularly high when materially deprived individuals experience another social disadvantage related, for example, to age, gender, physical condition or migrant/minority status. Experiencing multiple social vulnerabilities, especially when low-income and unemployed people are involved, can definitely aggravate the status of those transport-disadvantaged and raise the risk of social exclusion.

References

Barter, Paul A. 1999. "Transport and urban poverty in Asia. A brief introduction to the key issues." In *Regional Development Dialogue* 20, no. 1: 143–163. http://www.fukuoka.unhabitat.org/docs/occasional_papers/project_a/06/transport-barter-e.html. Accessed 12 June 2018.

European Commission. 2010. "Europe 2020. A strategy for smart, sustainable and inclusive growth" http://ec.europa.eu/eu2020/pdf/COMPLET%20EN%20BARROSO%20%20%20007%20-%20Europe%202020%20-%20EN%20version.pdf. Accessed 12 June 2018.

European Commission. 2013. "Attitudes of Europeans towards urban mobility." Special Eurobarometer 406/ Wave EB79.4. http://ec.europa.eu/commfrontoffice/publicopinion/archives/ebs/ebs_406_en.pdf. Accessed 13 June 2018.

European Commission. 2014. "Quality of transport." Report. Special Eurobarometer 422a/Wave EB82.2. http://ec.europa.eu/commfrontoffice/publicopinion/archives/ebs/ebs_422a_en.pdf. Accessed 13 June 2018.

European Commission. 2015. "EU Survey on issues related to transport and mobility." Joint Research Centre. http://publications.jrc.ec.europa.eu/repository/bitstream/JRC96151/jrc96151_final%20version%202nd%20correction.pdf. Accessed 13 June 2018.

Eurostat. 2016. "Statistics explained: People at risk of poverty or social exclusion." http://ec.europa.eu/eurostat/statistics-explained/index.php/People_at_risk_of_poverty_or_social_exclusion. Accessed 20 June 2018.

Eurostat. 2017. "Smarter, greener, more inclusive – indicators to support the Europe 2020 strategy." Luxembourg: Publications Office of the European Union. http://ec.europa.eu/eurostat/documents/3217494/8113874/KS-EZ-17-001-EN-N.pdf/c810af1c-0980-4a3b-bfdd-f6aa4d8a004e. Accessed 13 June 2018.

Eurostat. 2018a. "Statistics explained: Europe 2020 indicators – employment." http://ec.europa.eu/eurostat/statistics-explained/index.php?title=Europe_2020_indicators_-_employment, updated on 2018. Accessed 13 June 2018.

Eurostat. 2018b. "Euro area unemployment at 8.5%." https://ec.europa.eu/eurostat/documents/2995521/8782899/3-04042018-BP-EN.pdf/15f41da1-720e-429b-be25-80f7b2fb22cd. Accessed 10 June 2020.

Litman, Todd. 2003. "Social inclusion as a transport planning issue in Canada." Victoria Transport Policy Institute. https://www.researchgate.net/profile/Todd_Litman/publication/37183839_Social_Inclusion_as_a_transport_planning_issue_in_Canada/links/544a94ca0cf24b5d6c3ccb25.pdf. Accessed 20 June 2018.

Litman, Todd. 2017. "Transportation affordability. Evaluation and improvement strategies." Victoria Transport Policy Institute. https://www.vtpi.org/affordability.pdf. Accessed 20 September 2020.

Lucas, Karen, Giulio Mattioli, Ersilia Verlinghieri, and Alvaro Guzman. 2016. "Transport poverty and its adverse social consequences." *Proceedings of the Institution of Civil Engineers – Transport* 169, no. 6: 353–365. https://doi.org/10.1680/jtran.15.00073.

Mattioli, Giulio. 2017. "'Forced car ownership' in the UK and Germany. Socio-spatial patterns and potential economic stress impacts." In *Social Inclusion Regional and Urban Mobility: Contribution to Social Inclusion* 5, no. 4: 147–160. https://doi.org/10.17645/si.v5i4.1081.

Moore, José, Karen Lucas, and John Bates. 2013. "Social disadvantage and transport in the UK. A trip-based approach." https://www.tsu.ox.ac.uk/pubs/1063-moore-lucas-bates.pdf. Accessed 30 March 2020.

Titheridge, Helena, Nicola Christie, Roger Mackett, Daniel Oviedo Hernández, and Runing Ye. 2014. "Transport and poverty. A review of the evidence." https://www.ucl.ac.uk/transport/sites/transport/files/transport-poverty.pdf. Accessed 30 March 2020.

Transport for London. 2014. "Understanding the travel needs of London's diverse communities. A summary of existing research." http://content.tfl.gov.uk/understanding-the-travel-needs-of-london-diverse-communities.pdf. Accessed 13 June 2018.

9 Impacts on mobility in an ageing Europe

Vasco Reis and André Freitas

Abstract

An elderly person is a person aged 65 or over according to the definition of the European Union. At EU level, about one fifth of the population is elderly. Life expectancy has been increasing consistently in the last few decades. The impact of this trend on overall mobility demand is yet unclear, in part because elderly people are increasingly mobile. Even so, the natural ageing process is accompanied by a gradual deterioration in physical and psychological traits.

The increasing digitisation of mobility systems is another relevant aspect. The fast pace of technological development is known to exclude those less tech-savvy, which is common among elderly people. Finally, these demographic changes pose substantial challenges to authorities financing public transport services.

Introduction

There is no consensus regarding the definition of an elderly person and the World Health Organisation (WHO) stresses that it should not be regarded as a fully uniformed concept as it is westernised and may not fit some local realities, namely those related to poverty (WHO 2016). Nevertheless, the common definition accepted within the EU refers to an elderly person as a person aged 65 or over (young-old) and a very old person as a person aged 85 or over (old-old) (Eurostat 2017a, OECD 2018). The former group refers to those people who continue to have an active and independent life, normally after retirement. The latter refers to those less active people (due to some sort of disability, either cognitive or physical) and who require help from other people.

The following graph conceptualises mobility decline over the years, highlighting two segments of elderly people: young-old and old-old (Figure 9.1). The evolution from one to another segment is gradual and there is no determined age. A possible estimation can be obtained from the indicator of healthy life years at the age of 65. At EU level, this indicator is on average a decade. Hence, the segment of the young-old may range between 65 and 74 years and the segment of the old-old is after 75 years. A second important

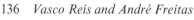

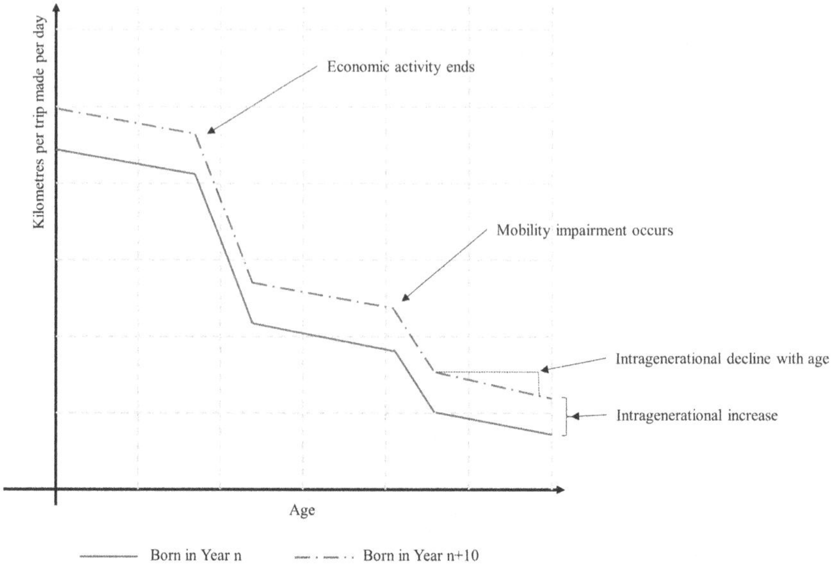

Figure 9.1 Conceptual model of mobility while ageing.
Source: Authors, based on Zmud et al. 2017.

insight from the graph is the representation of intergenerational mobility gains. The young generation of elderly people has more enhanced mobility than older generations, remaining active and mobile for more years. This is the consequence of improvements to social care services or the medical and health support system. Mobility systems must thus be planned to cope with the changing needs of these new generations of highly active and mobile elderly.

Overall figures from the EU-28 show that about one fifth of its population is aged 65 and above.[1] Furthermore, it is in rural regions that older people account for a higher proportion of the population (Eurostat 2017a). Most of the areas with high proportions of elderly persons are not only rural but sometimes also remote. Nevertheless, this reality is not so widespread as, for example, in some areas of Eastern Europe, namely Poland (Eurostat 2017a). Life expectancy has been increasing consistently in the last few decades and 2015 data indicate a life expectancy at birth of around 80 years, given that women have a life expectancy six years higher than men.

However, as life expectancy at birth is a fluctuating value, it is also interesting to consider life expectancy at the age of 65. This figure was estimated to be around two decades (Eurostat 2017a). The higher life expectancy for women seems to have an impact on the proportion of women living alone aged 65 and over, which is higher than men (Eurostat 2015). Globally, in 2013 elderly citizens represented 40% of single-person households in the EU

(Eurostat 2015). An important related aspect concerns the healthy life years at the age of 65. This influences several of the previously discussed characteristics, including the level of activity and mobility, employment or dependency ratios. According to Eurostat, on average a European has around a decade of healthy life after the age of 65. However, there are substantial variations among member states.

The growing level of access to digital tools and ICTs among older citizens should also be noted. More and more elderly people use the internet and digital technologies, both due to the natural aging of younger generations who have used the internet for quite some time or simply because people learn new skills. Even so, this segment remains somewhat wary about technology, particularly where computers and the internet are concerned. In 2016, almost half of the elderly population in the EU-28 used the internet at least once a week. In 2006, it was just a small proportion (Eurostat 2017a).

The increasing proportion of elderly people is one of the main challenges facing the EU in the next decade, both socially and economically. The growing technological capabilities need to be considered in any policy design, no matter what field. Policy planning and design for anything longer than the short term must consider the fast-growing digital capabilities gained from one generation to the next.

Mobility related characteristics

National populations are ageing as longevity increases. The impact of this trend on overall mobility demand is as yet unclear, in part because increasing longevity means that the characteristics of particular age cohorts change. By way of example, over time older women hold driving licenses, while frailty and loss of independent mobility tend to occur later in a longer life course (Metz 2013).

According to a recent IFMO study (Zmud et al. 2017), as the proportion of elderly increases, and the relative proportion of adults decreases, the effects both on demand and supply of transport can be considered relevant. Conversely, in rural and less developed areas, elderly people tend to be less mobile, due to a poor offer of transport services and less accessible transport infrastructure, coupled with physical and cognitive impairments.

Elderly people are increasingly mobile. Often people in their 60s (and even older) can be seen (and see themselves) as still very active, maintaining a very high overall activity level, traditionally not associated with their age. Financial resources, overall health and mobility do have an impact on those (self)perceptions (Samek Lodovici et al. 2012, Institute for Mobility Research 2013).

A higher life expectancy, better overall health and increased inclusion in the workforce are factors supporting this increased activity. Furthermore, factors like pensions, changing living arrangements and social connections combined with more varied and better transport options are related to the amount of travel and mode choices of older people (Institute for Mobility Research 2013). It can thus be perceived that there is growing transport demand from elderly people due to better health conditions, improved travelling solutions, more foreign-language skills and travelling lifestyles (Samek Lodovici et al. 2012). This trend may also lead to more cycling within this age group. For example, in Belgium, elderly people are early adopters of e-bikes (CIVITAS 2016).

Moreover, walking and cycling are seen across the EU as increasing activities among elderly citizens who wish to maintain active lives, as these are not only easy to maintain and accessible means of transport, but also associated with advantageous health outcomes (McDonald et al. 2013).

As elderly people are currently increasingly more active until later periods in their lives, it can be suggested that public transport may be crucial in maintaining active lifestyles even in cases where driving is no longer a possibility. Public transport is therefore very important as a support for older people's quality of life, improving their sense of freedom and autonomy (even more so in rural areas), guaranteeing access to basic services and decreasing social isolation (Shrestha et al. 2017).

Recognising the relevance of this age segment and its mobility, the European Commission funded the GOAL (Growing Older and staying mobile) project, within which five profiles for elderly people were defined – 'Fit as a Fiddle', 'Hole in the Heart', 'Happily Connected', an 'Oldie but a Goodie' and 'Care-Full' (based on data available through the SHARE database, survey of Health, Ageing and Retirement in Europe – http://www.share-project. org). Figure 9.2 shows the age and activity level of the profiles of older people as described in GOAL. These profiles vary in several characteristics, namely age, general health conditions, mental capabilities, reasons for travel, driving skills, need for assistance, among others (McDonald et al. 2012).

The scope of GOAL included the study of the requirements of public transport for older people, specifically addressing four main areas: affordability, availability, accessibility and acceptability.

Some segments of elderly people do experience mobility limitations caused by significant life-changing events such as increasing cognitive problems and physical impairments. The transition from using a car to using other transport modes will eventually occur for most people, namely for health or economic reasons or simply the responsibility of driving becomes too great. Nevertheless, such a transition will be very different according to personal conditions and experiences, and considering different social contexts (McDonald et al. 2013).

Moreover, travel behaviours and mobility patterns change as a person ages. According to the Mobilität in Deutschland survey (2008), cited by Hounsell and colleagues (Hounsell et al. 2016), ageing (after 55) is associated with more walking, less driving and more usage of public transport (especially after 75). Consequently, this segment of the population is associated with fewer journeys when compared to younger adults and will likely

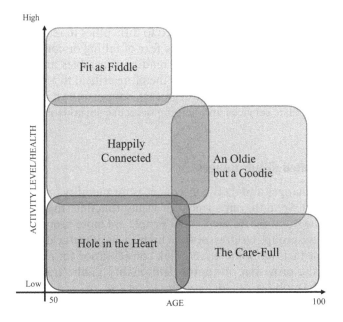

Figure 9.2 Age and activity level of the profiles of older people.
Source: Authors, based on McDonald et al. 2013.

change their transport mode (Hounsell et al. 2016). Notwithstanding, the preferences of the elderly are changing. By way of example, in the United Kingdom, the option of walking has been decreasing for some decades (McDonald et al. 2012).

Regardless, the mobility of older people is likely to be dependent on an adequate supply and appropriate quality of public transport services (Hounsell et al. 2016). Mollenkopf and Flaschenträger (2001, cited in Hounsell et al. 2016) found that "almost all older persons, regardless whether they participate in walking, cycling, driving or using public transport, suffer from the tighter and more aggressive traffic".

The reduction in travel in older age groups may also simply arise as a result of their smaller presence in the workforce. Therefore, older Europeans tend to use urban public transport mostly for leisure activities (shopping, visiting friends and relatives) to take children to school and to other after-school activities and to access healthcare services. The preservation of such activities is related to the availability of public transport, which is therefore of importance for the quality of life of the elderly (McDonald et al. 2013).

Often, changes in mobility patterns will be related to the increasing difficulty in overcoming different barriers that might occur due to the ageing process. Older people may face physical, psychological and economic barriers to travel. These may include diminished motor, sensory and cognitive abilities (ECMT 2002). For example, regarding public transport, transport-related barriers can be linked to difficulties in reaching bus stops or getting in and out of vehicles to the fear of falling or concerns with personal security, or even difficulties in reading timetables and destinations. Improvements in public transport are therefore critical to an "age-friendly" approach, especially among rural segments, supporting an independent life and access to basic services and helping decrease social isolation (Hounsell et al. 2016).

Mobility-related disadvantages

Age-related changes and their consequences for mobility include decreased flexibility and strength, impairment of visual function, increased vulnerability to bone fracture. As such, older people are more prone to be affected than other age groups by stressors like high levels of traffic density and flow, as well as the fact that some drivers lack consideration for other road users. Therefore, the provision of appropriate quality public transport is paramount for the mobility of older people.

It is well known that the elderly population has significantly increased its average mileage per year and has a high motorisation rate. There has been significant growth in kilometres travelled per day by seniors in the study countries (Institute for Mobility Research 2013):

- 70% increase in England from 1982 to 2012;
- 40% increase in Germany from 1982 to 2012;
- 40% increase in the United States from 1983 to 2008; and
- 30% increase in Japan from 1987 to 2010.

Nevertheless, for each additional year of age, senior mobility declines overall at about one kilometre per person per day. In such a context, driver attitude and driving behaviour are some of the factors that can influence this segment's transition to public transport (Shrestha et al. 2017).

Overall, whilst mobility indisputably declines with age, successive generations are nevertheless starting their declines at higher levels of mobility, for which the main contributor is car ownership. In Germany, for example, the percentage of elderly people owning a car has tripled since the mid-1980s – more than for any other age group (Institute for Mobility Research 2013). But whilst car ownership rates for older people has increased, car use actually decreases with age, probably due to an increasingly challenging driving environment.

These changes may also be connected with travelling for tourism purposes. Elderly people represented around one fifth of EU tourists, which is still lower than the share of the EU population that this segment represents (Eurostat 2017b). On average, almost half of EU elderly people travel for tourism purposes and above one quarter travel to another EU member state. Elderly people tend not to be employed and so are able to benefit from discount and budget trips offers. Data on this provide some significant consequences in terms of externalities. Safety issues and accident rates (report accident rates by age group) in relation to declining driving capabilities are perceived as one of the reasons not to drive. In fact, the highest proportion of accidents involving older drivers and for which they are responsible is somehow related to perception and decision-making issues (Verhaegen 1995). The GOAL project identified the causes of accidents with physical injury involving elderly people in the old-old segment (Figure 9.3). It can be seen that the use of the wrong way, together with other driving errors account for more than one third of cases. Moreover, European accident data show that older car occupants, pedestrians and cyclists have significantly higher risks of severe and fatal injuries. Male elderly citizens seem to be particularly at risk when it comes to cycling accidents, while women have an increased risk as pedestrians (Wisch et al. 2017).

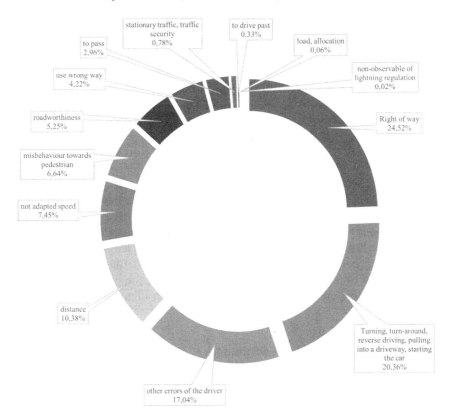

Figure 9.3 Causes of traffic accidents with physical injury for age group 75 and older.

Source: Authors, based on McDonald et al. 2012.

Despite the safety-related issues, high levels of car ownership can make transitions to other means of transportation that could somewhat compensate for increasing cognitive problems and physical impairments difficult. In fact, some elderly people who have previously relied mostly on their car can find it very challenging to transition from driving to using public transport (Shrestha et al. 2017).

Independently of the mode, accessibility is a key issue for older people using transport systems. Regarding public transport, relevant issues are the location of bus stops in relation to the trip origin and destination(s), the quality of the infrastructure for the walking sections of the journey and the accessibility of the vehicles themselves (e.g. whether they feature a low floor). Even in the light of recent progress improving accessibility for previously excluded segments of the population within the EU, about "10 to 20% of European citizens (namely people with disabilities and older people) still face barriers and reduced accessibility to transportation" (Shrestha et al. 2017, 347). For older people, it is essential to guarantee the provision of public transport within the reach of people's origins and destinations, with adequate service times and frequencies, which may assume the form of special transport services if user experience reduced physical abilities.

The reluctance that some older people may show in using public transport may be related to their health, but also because of difficulties caused by the system itself. In fact, those who used public transport their entire lives are usually more prone to using different transport alternatives. On the contrary, those who mostly relied on their own car as their main transport mode tend to see public transport as complicated and inconvenient, mostly due to their lack of previous experience (Adler and Rottunda 2006).

Considering the potential impacts of injuries for older people, safety is a serious concern for this age group. The likelihood of a longer recovery period and a greater psychological impact than a younger person in a similar incident play a significant role in this perception. These may be the reasons why older people worry about their safety and are reluctant to take public transport or use bikes, for example, together with the fear of crime or falling over and becoming injured (Shrestha et al. 2017). On the other hand, slow journey times might not be seen as a barrier for older people when compared with other segments of the population (Transport for London 2014).

Multiple socio-economic disadvantages

Elderly people have more critical socio-economic characteristics than younger generations, owing to their position in the labour market and health conditions. The socio-economic characteristics of this group should be considered with particular attention on several factors, most of which are interrelated: household income, working status and disability or impairment affecting travel. Household income can be strongly influenced by retirement (which in many cases represents a lower net monthly income), loss of a spouse or illness, just to name a few.

The percentage of older people living alone in the EU (almost one third) highlights some of the increased susceptibility of this segment, as it represents not only lower incomes, but, most likely, isolation. Moreover, the percentage of older people living alone may represent a strong disadvantage for this segment's mobility (Eurostat 2015).

Mobility impairments are also associated with declines in mobility. Despite increasing life expectancy, this does not necessarily represent an extended quality of life, especially in light of the increasing prevalence of stressors such as smoking, diabetes, obesity and low levels of exercise. These conditions can significantly increase the need to allocate a greater share of household spending to rising healthcare costs, contributing to a reduction in discretionary household income and dwindling wealth accumulation.

The risk of poverty among elderly people is associated with decreased mobility and restrictions on access to transport. These older citizens will make significantly fewer trips and cover less distance daily than people with higher incomes. For example, for disadvantaged older women in low-status residential areas, trips will mostly cover their local residential environment (Giesel and Köhler 2015).

Hence, affordability is also a relevant topic for many elderly people, especially for those with less available income in retirement. In a context of more limited resources, the cost of travelling will become a major barrier for many old people to travel as often as they would like. In extreme cases, the cost of transport can represent a barrier to access basic and necessary services (hospitals, supermarkets, pharmacies) (Shrestha et al. 2017). For older people in such situations, who usually have more time and less money, travel costs become more important, leading to the choice of cheaper alternatives that require longer travel times.

Conclusions

Elderly people's characteristics will keep evolving considerably fast. It is likely, for example, that activity levels among elderly people will increase in the future. Life expectancy will continue to increase and it can be expected that senior people will remain employed for longer, taking on second careers or volunteer activities. These changes will have an impact both in economic and psychological terms. Nevertheless, ageing cannot be stopped and senior people will, sooner or later, face physical difficulties that will be accompanied by certain inevitabilities. Therefore, it is possible at the same time to find factors that support as well as hinder mobility patterns alongside the ageing process. It must be recognised that today's seniors are in fact a very diverse population segment and that therefore not all people will react in the same way. Furthermore, as societies evolve, so do mobility and travel patterns, associated with a high level of uncertainty.

Another relevant aspect is that mobility systems must ensure digital inclusion. The fast pace of technological development is known to exclude those less tech-savvy, which is common among elderly people, notably in the group of the old-old. This situation is paradoxical in the sense that new mobility solutions, such as autonomous vehicles or ride hailing services, can help precisely those with mobility impairments, which appear once people age. Yet, due to lower levels of digital literacy, those who could benefit the most are at risk of being kept at bay. Such a vicious circle may be, and is being, interrupted when engaging the social networks of senior citizens. Younger people (e.g. relatives, friends, neighbours) can assist the elderly in using new mobility solutions. The benefits should largely go beyond the field of mobility, as they also contribute to nurturing the social network of the elderly, which is at risk of decaying over time.

Demographic changes already pose substantial challenges to authorities financing public transport services and the pressure to find adequate solutions will further increase. There is still a research gap regarding such solutions. In fact, in the vast majority of countries, elderly people have access to several travel discounts. In some cases, as in the UK, retired people are granted free bus travel, with only minor restrictions usually related to peak periods.

As the population pyramid gets inverted, the share of subsidised public transport users is likely to increase, whereas passengers paying the full price will diminish, raising new funding challenges that will have to be dealt with in the short term by the relevant stakeholders. This phenomenon is even more striking in rural areas where the share of older people is higher than in urban areas.

Note

1 It should however be mentioned that such proportions do vary significantly across Member States. In 2016, the three highest shares were found in the central Greek region of Evrytania (30.7%), the north-western Spanish region of Ourense (30.7%) and the West Flanders municipality of Veurne in Belgium (30.2%) (Eurostat 2017a).

References

Adler, Geri, and Susan Rottunda. 2006. "Older adults' perspectives on driving cessation." In *Journal of Aging Studies* 20, no. 3, 227–235. https://doi.org/10.1016/j.jaging.2005.09.003.

CIVITAS. 2016. "CIVITAS thematic policy note – transport poverty." https://civitas.eu/sites/default/files/civitas_policy_note_transport_poverty.pdf. Accessed 2 April 2020.

ECMT. 2002. "Transport and ageing of the population." European Conference of Ministry of Transport. https://read.oecd-ilibrary.org/transport/transport-and-ageing-of-the-population_9789264187733-en#page12. Accessed 2 April 2020.

Eurostat. 2015. "People in the EU. Who are we and how do we live?" https://ec.europa.eu/eurostat/statistics-explained/index.php/People_in_the_EU_%E2%80%93_who_are_we_and_how_do_we_live%3F. Accessed 2 April 2020.

Eurostat. 2017a. "People in the EU – statistics on an ageing society – statistics explained." http://ec.europa.eu/eurostat/statistics-explained/index.php?title=People_in_the_EU_-_statistics_on_an_ageing_society. Accessed 2 April 2020.

Eurostat. 2017b. "Tourism statistics." http://ec.europa.eu/eurostat/statistics-explained/index.php/Tourism_statistics_-_participation_in_tourism#Participation_in_tourism_was_significantly_lower_among_persons_aged_65_and_over. Accessed 20 June 2018.

Giesel, Flemming, and Katja Köhler. 2015. "How poverty restricts elderly Germans' everyday travel." In *European Transport Research Review* 7, no. 15, 1–9. https://doi.org/10.1007/s12544-015-0164-6.

Hounsell, Nick B., Birendra P. Shrestha, Mike McDonald, and Alan Wong. 2016. "Open data and the needs of older people for public transport information." In *Transportation Research Procedia* 14 (January). 4334–4343. https://doi.org/10.1016/J.TRPRO.2016.05.355.

Institute for Mobility Research. 2013. *Mobility Y. The Emerging Travel Patterns of Generation Y. Institute for Mobility Research.* Munich. https://www.ifmo.de/files/publications_content/2013/ifmo_2013_Mobility_Y_en.pdf. Accessed 13 June 2018.

McDonald, Mike, Nick Hounsell, Alan Wong, Birendra Shrestha, Niccolò Baldanzini, Avinash P. Penumaka, and Ingrid Hendriksen. 2012. GOAL project. European Commission. https://www.southampton.ac.uk/engineering/research/projects/goal_growing_older_staying_mobile.page#project_overview%0A. Accessed 2 April 2020.

McDonald, Mike, Nick Hounsell, Alan Wong, Birendra Shrestha, Niccolò Baldanzini, Avinash P. Penumaka, and Ingrid Hendriksen. 2013. "Transport needs for an ageing society – action plan." European Commission. https://trimis.ec.europa.eu/sites/default/files/project/documents/20140115_095617_32515_Action_Plan__Transport_Needs_of_an_Aging_Society.pdf. Accessed 2 April 2020.

Metz, David. 2013. "Peak car and beyond. The fourth era of travel." In *Transport Reviews* 33, no. 3, 255–270. https://doi.org/10.1080/01441647.2013.800615.

OECD. 2018. "Elderly population (indicator)." https://data.oecd.org/pop/elderly-population.htm#indicator-chart. Accessed 2 April 2020.

Samek Lodovici, Manuela, Flavia Pesce, Patrizia Malgeri, Silvia Maffii, and Caterina Rosa. 2012. "The role of women in the Green Economy – the issue of mobility." Policy Department C: Citizens' Rights and Constitutional Affairs – European Parliament. Brussels. http://www.europarl.europa.eu/RegData/etudes/note/join/2012/462453/IPOL-FEMM_NT(2012)462453_EN.pdf. Accessed 14 June 2018.

Shrestha, Birendra P., Alexandra Millonig, Nick B. Hounsell, and Mike McDonald. 2017. "Review of public transport needs of older people in European context." In *Journal of Population Ageing* 10, no 4: 343–361. https://doi.org/10.1007/s12062-016-9168-9.

Transport for London. 2014. "Understanding the travel needs of London's diverse communities. A summary of existing research." http://content.tfl.gov.uk/understanding-the-travel-needs-of-london-diverse-communities.pdf. Accessed 13 June 2018.

Verhaegen, Paul. 1995. "Liability of older drivers in collisions." In *Ergonomics* 38, no 3, 499–507. https://doi.org/10.1080/00140139508925121

WHO. 2016. "Definition of an older person. Proposed working definition of an older person in Africa for the MDS Project." In *World Health Organisation*. https://www.who.int/healthinfo/survey/ageingdefnolder/en/. Accessed 2 April 2020.

Wisch, Marcus, Elvir Vukovic, Roland Schäfer, David Hynd, Adam Barrow, Rahul Khatry et al. 2017. "Road traffic accidents involving the elderly and obese people in Europe incl. investigation of the risk of injury and disabilities." http://www.seniors-project.eu/download/public-files/public-deliverables/SENIORS_D1.2_Crash_data_Hospital_statistics_DRAFT.pdf. Accessed 2 April 2020.

Zmud, Johanna, Lisa Green, Tobias Kuhnimhof, Scott Le Vine, John Polak, and Peter Phleps. 2017. "Still going ... and going: The emerging travel patterns of older adults." https://www.bmwgroup.com/content/dam/grpw/websites/bmwgroup_com/company/downloads/en/2017/2017-BMW-Group-IFMO-Publication.September.pdf. Accessed 2 April 2020.

10 The predicaments of European disabled people

Vasco Reis and André Freitas

Abstract

In 2012, around 59 million Europeans (aged 15 and over) reported a disability with regards to mobility or transport. Disabled people have specific mobility problems depending on the cause of their impairment (e.g. reduced vision, hearing or movement, environmentally or psychologically challenged).

There is now wide recognition of the importance of issues such as access to transport and the impact that it can have on the quality of life and independence of disabled people. Yet, there is no general agreement nor clear understanding about most disabled population mobility habits. People with a disability are a very heterogeneous group with several different types of impairment, which inhibit their travel options in different ways and consequently their personal quality of life and independence.

Conceptual discussion

On 13 December 2006, the Convention on the Rights of Persons with Disabilities and its Optional Protocol was adopted. The negotiation took place between 2002 and 2006, making it the fastest negotiated human rights treaty.[1] At the time, a central aspect of the Convention was to raise the cultural and social position of disabled people from *objects* or recipients of charity, social protection or medical treatment to (human) *subjects* with rights. As pointed out by Cuthbertson (2015), the prevalent symbol of human power and privilege is of a walking unaided person. Other forms of locomotion (e.g. a wheelchair) carry strong negative cultural meanings.

The negative connotation of disability is grounded in deep-rooted stereotypes, myths and ideologies, nurtured over time. Oliver (1990) argued that disabled people are regarded as problems since they deviate from the dominant culture's view of what is expected, normal or socially accepted. Literature clearly indicates that disabled people face social and spatial exclusion. Disabled people have a higher probability of living in poor neighbourhoods, with inadequate access to transport, equipment, services or employment (Gleeson 2006). By way of example, even nowadays, many underground stations are inaccessible to wheelchairs. Disabled people are often trapped in a self-reinforcing vicious circle of poverty and isolation.

Yet, disability is not limited to locomotion. Urry (2007) identified five forms of interdependent mobility and, hence, of disability: corporeal, imaginative, virtual, communicative and mobility of objects (circulatory and logistical). Communication is of particular relevance. The recent technological developments, notably regarding Information and Communication Technologies, brought an array of new technology-driven mobility services such as ride hailing (e.g. UBER), micro-mobility services and paperless ticketing systems. These new services have been designed for non-disabled people. Disabled people may find themselves even more excluded from the mobility system. Such a situation is, in itself, paradoxical. Newer regimes and ideologies, such as neoliberalism, should have brought unprecedented freedom, freeing disabled people from the various forms of immobilities. Instead, they are creating new forms of exclusion and injustice, further aggravating their already inferior position. The day-to-day reality of disabled people is of restricted mobility, immobility or a continuum of situations that serve to highlight their impairment and inferiority (Imrie 2000).

Urbanisation seems to play a relevant role in the level of exclusion. Peri-urban and rural areas exhibit distinctive organisation, dynamics and features of mobilities. Services tend to be located further away and mobility services are less abundant and accessibility is also inferior. Specialised services for disabled people are scarce or, when available, must be requested in advance.

Quantitative assessment and characteristics

A disabled person or a person with reduced mobility is usually considered to be any given person whose mobility while using a mode of transport is reduced because of physical disabilities (sensory or affecting mobility, being permanent or temporary), intellectual impairment or any other cause of disability, or age, which requires appropriate and specific attention as well as an adaptation of the transport service made available to all passengers and all their particular needs.

Long-term physical, mental, intellectual or sensory impairment in its interaction with different types of transport-related barriers may affect their full and effective participation in society on an equal basis with others (United Nations 2008). Nonetheless, there are multiple dimensions that can prevent people from performing one or several basic activities.

The definitions that tend to be applied for statistical purposes depend mostly on the number of questions that can be asked about issues such as impairments, limitations or barriers to participation. Table 10.1 details the differences in data sources for EU statistics when addressing the topic of disability.

Table 10.1 Overview of data sources for EU statistics on disability

	European health and social integration survey (EHSIS)	European health interview survey (EHIS)	Statistics on income and living conditions (SILC)	Ad hoc module on employment of disabled people in the labour force survey (LFS)
Main topics covered	Disability as defined by the UN Convention	Health status, health determinants and health care use	Income, social inclusion and living conditions	Employment of disabled people
Legal basis	No	Yes	Yes	Yes
Periodicity	Once (2012)	5-yearly	Annual	Irregular (2002 and 2011)
Limitations in usual activities caused by a health problem	Yes	Yes	Yes	
Difficulties in carrying out basic activities	Yes	Yes		Yes
Difficulties in performing personal care activities	Yes	Yes (persons aged 65+)		
Difficulties in performing household care activities	Yes	Yes (persons aged 65+)		
Participation restriction linked to a health condition and/ or a basic activity difficulty	Yes			
Limitation in work caused by health problems and/ or difficulties in basic activities	Yes			Yes

Source: Eurostat (2018).

Despite the existence of a significant number of questions related to impairments, limitations and barriers to participation[2], we can see in Table 10.2 the distribution of all EU citizens (aged 15 and over) who report a disability by categories of life areas in which this disability is a source of constraint and by European country.

In 2012, 70 million people reported disabilities (aged 15 and over) in the EU-27. This number does not mean that all these persons have mobility problems. Also, in 2012, around 44.5 million people reported some sort of difficulty in conducting basic activities (including mobility) in the EU. In the same year, 52.9% (37.03 million) and 31.7% (22.19 million) of the total EU population (aged 15 and over) reported a disability in mobility and transport life areas, respectively (Eurostat 2018).

Table 10.2 Share of disabled persons aged 15 and over reporting a disability in the specified life areas, 2012 (as a percentage of persons reporting a disability in at least one area)

	Life areas									
	Mobility	*Transport*	*Accessing buildings*	*Education and training*	*Employment*	*Using the internet*[a]	*Social contact*[a]	*Leisure pursuits*	*Paying for the essential things in life*	*Perceived discrimination*
EU-27	52.9	31.7	37.0	25.6	38.6	4.6	2.0	60.9	22.7	19.8
Belgium	46.4	24.5	31.2	27.9	44.3	3.3		64.1	11.1	23.3
Bulgaria	44.6	34.7	33.2	9.8	26.6	2.0	2.4	35.2	76.4	9.6
Czech Republic	58.1	39.7	40.5	13.9	35.1	2.1	2.4	67.4	34.4	18.5
Denmark	51.0	22.1	33.7	33.6	51.5	7.6	2.8	67.6	11.9	24.3
Germany	50.4	20.8	33.4	31.6	37.9	4.2	1.1	66.0	10.1	20.3
Estonia	56.6	27.1	28.9	14.7	24.7	2.5		55.0	44.2	11.4
Ireland	:	:	:	:	:	:	:	:	:	:
Greece	61.2	39.5	40.9	7.7	18.4	9.1	1.7	48.1	52.9	9.7
Spain	58.5	34.0	34.9	22.3	40.8	5.6	2.0	69.0	8.7	16.5
France	44.6	29.2	31.5	35.0	35.3			66.4	10.5	23.0
Croatia	:	:	:	:	:	:	:	:	:	:
Italy	66.2	51.4	42.7	12.3	20.1	1.7	2.3	54.4	35.2	12.1
Cyprus	49.6	39.2	32.6	19.6	33.3	3.9		52.2	35.6	15.1
Latvia	46.7	30.3	24.6	15.1	41.2	2.4	4.1	47.2	51.3	9.8
Lithuania	47.1	32.7	34.3	19.5	42.3	4.2	2.5	48.2	39.4	12.9
Luxembourg	40.7	15.1	26.4	26.9	34.0	5.1		63.8		23.4
Hungary	60.8	44.6	44.7	16.6	42.4	2.6		47.5	54.8	17.5
Malta[b]	48.7	30.4	34.4	16.6	6.7			43.0	43.5	9.0
Netherlands	36.2	22.3	25.3	25.7	47.1	2.5	2.1	62.7	12.2	21.4
Austria	39.9	20.1	27.7	21.0	36.2	2.0		55.4	12.2	14.8
Poland	51.0	35.8	38.5	28.5	43.5	2.6		48.9	37.2	17.3
Portugal	47.7	33.5	35.6	13.4	38.0			42.4	36.3	11.3
Romania	66.1	42.3	51.4	15.3	31.7	1.6	4.3	47.2	65.0	13.4
Slovenia	51.5	33.4	41.0	22.5	23.5		4.6	60.9	24.1	13.1
Slovakia	42.5	20.5	27.8	21.0	21.5	2.2		55.5	34.5	15.7
Finland	40.3	13.3	25.3	19.5	44.6	4.0		64.5	11.0	18.6
Sweden[c]	26.6	15.9	18.5	20.5	46.7	4.3	2.1	67.2	5.1	24.9
United Kingdom	55.7	31.4	45.1	35.7	54.6	9.4	3.6	70.3	10.9	33.3
Iceland	32.1	15.0	14.5	19.6	40.7			55.1	16.9	27.7
Norway[c]	44.4	15.6	27.0	37.0	66.0	7.2	2.8	65.1	4.9	27.0

Source: Eurostat (2018).

a Data with low reliability for most Member States and non-member countries.
b Barriers to employment and barriers of perceived discrimination: low reliability.
c Barriers to paying for the essential things in life: low reliability.

It was in two life areas where more than half of all people with disabilities (in the EU-27) reported that their disability was the cause of their restriction on participation in 2012: leisure pursuits (in other words, hobbies or interests that involve spending time with other people) and mobility (defined here as the ability to leave one's own home).

Table 10.2 confirms these arguments in all the EU member states and also covers other aspects like persons reporting a disability in education and training or in social contacts, for example. It is worth mentioning that, in that same year, women reported higher levels of disability regarding mobility (54% in women vs. 48% in men) and regarding transport (33% in women vs. 28% in men) alike.

Analysing all the barriers to participation that people with disabilities face recalls some of the demographic categories in that we can find higher or lower prevalence of disability. So, this prevalence of disability was higher for women (19.9%) than for men (15.1%) as evident in Figure 10.1 shown below (Eurostat 2018). Disabled women seemed more likely to report barriers to mobility, transport and to the accessibility of buildings than disabled men. The prevalence of disability was also much higher for people aged 65 and over (35.6%) than for those aged 45–54 (18.8%) or aged 15–44 (8.5%), as confirmed in Figure 10.1.

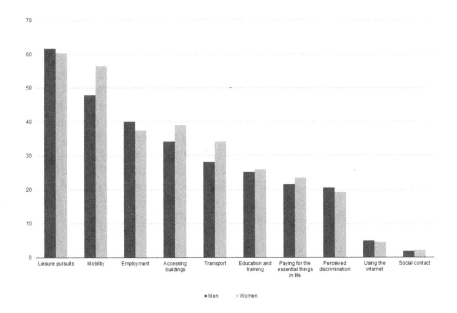

(*) Estimates
Source: Eurostat (online data code: hlth_dsi090)

Figure 10.1 Share of disabled persons aged 15 and over reporting a disability in the specified life areas, by gender, EU-27, 2012 (estimates) (as a percentage of persons reporting a disability in at least one area).
Source: Eurostat 2018.

Mobility-related characteristics

For the Europe 2020 strategy to be successful, the full economic and social participation of people with disabilities is essential.

Research has shown that disabled people travel less than non-disabled people. In a highly dense and urbanised context, for example, the public authority Transport for London (Transport for London 2014) has realised that whilst disabled people fundamentally have a similar modal share as non-disabled Londoners, the frequency of using transport is much lower among disabled persons: 1.97 journeys per weekday among disabled Londoners vs. 2.77 for non-disabled Londoners. "Public transport types are also less commonly used by disabled Londoners than non-disabled Londoners, 60 per cent of disabled Londoners have used any public transport (excluding walking) in the last year compared with 73 per cent of non-disabled" (Transport for London 2014, 196). Accessibility-related barriers top-rank the main obstacles that Londoner disabled persons encounter whilst using public transport. Such transportation obstacles allow disabled people even fewer opportunities to interact with their communities, thus enhancing social exclusion in what can be regarded as a vicious circle which is intensified in the presence of combined social layer groups (e.g. elderly people with a disability).

A study in England and Wales aimed to identify the attitudes of disabled people to public transport in the respective regions (DPTAC 2002, 14), making an effort to represent more effectively the broad range of disabled people when preparing advice to government. This study mentions that, when asked unprompted about their local concerns, transport issues are top of disabled people's list (48%), followed by crime (22%), environment (16%) and social services/facilities/community (16%). Transport issues for disabled people can be understood as: (i) the inaccessibility of public transport, where provision often fails to meet the diverse needs of young wheelchair users (transport disability); (ii) the importance of emotion in experiences of transport and the anxieties produced by inaccessible transport; and (iii) the centrality of private forms of transport in accessing leisure (mobility dependency) (Pyer and Tucker 2014, 38).

Specific transport concerns include (DPTAC 2002):

- Difficulty in using public transport (16%);
- Frequency of public transport (16%);
- Unreliable buses/trains (10%);
- Traffic congestion (8%);
- The speed of motorists (8%);
- Shortage of car parking (7%);
- Traffic noise (3%); and
- The level of road accidents (2%).

With the exception of the difficulty felt while using transport, these issues are similar to the general public. Hence, it is possible to assume that people with reduced mobility have special needs because of their physical or psychological limitations, but they have similar needs as non-disabled persons as well. Comparable results were found in more recent research projects such as the one carried out in 2014 by Birgitta Thorslund about the mobility behaviour among people with one of the most frequent sensory deficits in humans, hearing loss. These results show that "a higher degree of hearing loss was associated with less likelihood of having a driver's license. However, individuals with hearing loss who had a driver's license, drove as much as normal hearing drivers" (Thorslund 2014, 28). From this study, it was concluded that hearing loss is associated with higher use of private transport, the car being perceived as a "compensational tool for functional limitations" (Thorslund 2014, 55). However, it has no effect on the distribution of how much each type of transportation was used.

In contrast, Canadian-based data show a dissimilar trend. In this respect, a study shows that disabled Canadians travel considerably less and over shorter distances and have less access to key services than the average Canadian population (Paez et al. 2009). Karen Lucas, referring to the Canadian study, therefore argues that disability impacts very negatively on the well-being of disabled persons, reflected in the fact that they perform fewer trips than the non-disabled population. Their social lives are therefore hindered by limited access to transportation (Lucas 2012).

American researcher Bascom (Bascom 2017) examined how individuals with disabilities are meeting their transportation needs. He hypothesised (and effectively concluded) that individuals with disabilities who have stronger and wider social networks are more likely to ride-share and have access to other forms of transportation assistance than those who have weaker social networks, who will be much more likely to rely on public transportation. This is a concrete reflection of the widely popular sociological network theory of interpersonal ties, developed mostly by American sociologists in the 1970s.

Mobility-related disadvantages

Several scholars have related spatial and social inequalities in access to transport for particular social groups (Kenyon et al. 2002, Preston and Rajé 2007). There is now wide recognition of the importance of issues such as access to transport and the impact that it can have on the quality of life and independence of people with disabilities, as they have specific mobility problems. The characterisation of the mobility disadvantages in this segment is particularly difficult, due to the wide diversity of disabilities or impairments. As previously discussed, the European Parliament defined five types of disabilities (European Parliament and the Council of the European

Union 2006). What follows is a brief identification of the key mobility disadvantages per type:

1   Reduced vision (vision impaired) – key challenges include: situational awareness, wayfinding in terminals, acquisition of tickets and understanding any visual-based information;
2   Reduced hearing (hard of hearing) – key challenges include: understanding any sound-based information, which is of particular relevance in emergency situations, or even detecting any risky situation.
3   Reduced movement (mobility impaired) – key challenges are linked with the need to overcome different heights (e.g. different levels of the terminal, entering or exiting vehicles) and to overcome gaps (e.g. between the terminal platform and vehicle).
4   Environmentally challenged (allergic) – key challenges are related to a higher-than-average concentration of pollutants in or around vehicles and terminals. Several vehicles are powered by internal combustion engines (e.g. buses, taxis, aircraft, ships). In addition, in/around terminals (e.g. airports, bus terminals), vehicles are frequently involved in manoeuvres or move at low speeds. These are two situations where internal combustion engines are the least efficient (producing the highest levels of emissions). Moreover, certain terminals and/or routes (e.g. metropolitan) are covered, which precludes efficient air circulation and favours the concentration and deposition of pollutants. People with health conditions may be particularly affected in these areas.
5   Psychologically/mentally cognitively challenged – a key challenge is related to the ability of the person to understand how to use the transport system, including knowing what ticket to buy, wayfinding in the terminal and situation awareness.

People with reduced mobility are less likely to benefit from access to standard means of transport if the initial design does not take their needs into account. Hence, persons with a disability tend to rely on private transport to access services and for day-to-day activities such as shopping and participating in social activities. In fact, the single most frequently used mode of transport by people with reduced mobility is the car as passenger (DPTAC 2002).

Arguably, in order to realise many opportunities for disabled people to participate fully in society, it is common for them to depend on the support of relatives who chauffeur them by private transport or accompany them on public transport.

Pyer and Tucker (2014) conducted an ethnographic investigation focused on teenagers in wheelchairs. They concluded that the main symptom of mobility poverty that affects this group results from inaccessibility to public transport vehicles, which seems to be an additional reason for applying the concept of "forced car ownership". This trend is consistent with data from the UK Department for Transport, mentioned by these authors, which point out that car ownership among a household of disabled people is well above the national average for families with dependent children. Even though not representative of each single type of impairment, the main conclusion that arises from this investigation is therefore that "the availability of private cars enabled access to a range of leisure spaces which would otherwise have been closed-down to many if they had been solely reliant on public transport" (Pyer and Tucker 2014, 48). The findings of Pyer and Tucker (2014) about the main source of exclusion are perfectly aligned with the ones from the Directorate-General for Internal Policies (Samek Lodovici and Torchio 2015), which highlight physical barriers and most notably public transport vehicle design as the main issue for the exclusion of people with reduced mobility.

To counteract these numbers and increase the number of people using public transport, national governments are introducing several new policy frameworks that intend, on the one hand, to contribute to the deployment of large infrastructure enhancements, conveying inclusive layouts for example and, on the other hand, to offer concessionary bus travel. In general, the take-up of concessionary passes is high, as demonstrated in this UK example: "older and disabled concessionary pass holders collectively make around 1.2 billion bus journeys, accounting for almost one in four of all journeys on local bus services" (Greener Journeys 2014, 4)[3].

It is simultaneously important to keep in mind that enhancements required for the benefit of disabled persons also favour those who do not suffer from any transport impairment. Buses featuring a low floor or low entry were meant to provide easy access for wheelchairs but have a positive indirect effect on the boarding and alighting time of all public transport users. The gist of accessible environments is creatively captured by the following image (Figure 10.2).

To combine transport infrastructures with social inclusion layouts and policies is even more important if one assumes that the number of citizens with disabilities and/or functional limitations is likely to increase significantly with the ageing of the European Union's population. It is equally important to keep pace with the tourism-induced indirect economic effects. In general, the accessible tourism industry, which statistics show is flourishing (Bowtell 2015), affects a wider scale of economy through the so-called "multiplicator effects" (Rebstock 2017).

Multiple socio-economic disadvantages

In the European statistics on income and living conditions (also known as EU-SILC[4]), disability is narrowed according to the concept of a global activity limitation, which it defines as a "limitation in activities people usually do because of health problems for at least the past six months" (Eurostat 2018).

Figure 10.2 Accessibility for all.
Source: Authors, based on Rebstock 2017, 6.

In 2013, according to the indicator "at risk of poverty or social exclusion" (AROPE), about 30% of the population aged 16 or more in the EU-28 and having an activity limitation was at risk of poverty or social exclusion, compared with 22% of those with no limitation. Similar results were obtained for the at-risk-of-poverty rate (19% vs. 15%), severe material deprivation rate (13% vs. 8%) and the share of individuals aged less than 60 and living in households with very low work intensity (24% vs. 8%). It is important to note that significant differences across member states are visible, yet in all of them people without activity limitation are on average less exposed to the risk of poverty and social exclusion than those with some activity limitation (Eurostat 2015).

The prevalence of disability in the previous year (2012) was higher for people having completed at most lower secondary education (25.0%) than for those having completed at most upper secondary or post-secondary, non-tertiary education (15.4%) or tertiary education (11.0%). Figure 10.3 may be a crucial support of such findings. It is also higher for retired persons

(34.3%) than for unemployed people (20.5%), other economically inactive people (20.2%) or employed persons (8.0%).

People with activity limitation have to rely heavily on social transfers. At the EU-28 level, 68% of the population aged 16 and over with some sort of disability would have been at risk of poverty if social transfers (e.g. social benefits, allowances and pensions) had not taken place. On the other hand, 31% of the population with no activity limitation would have been at risk of poverty.

Another piece of relevant statistics is reported in Figure 10.3. People with disabilities having completed tertiary education were less likely to report a disability for mobility than other people with disabilities with lower levels of education. Less than 1 person out of 2 with basic activity difficulties was employed. The employment rate of people with basic activity difficulties in the EU-28 in 2011 was around 47.3%.

One of the very few pieces of research about the relationship between disability and internet usage is noteworthy. This is a recent investigation

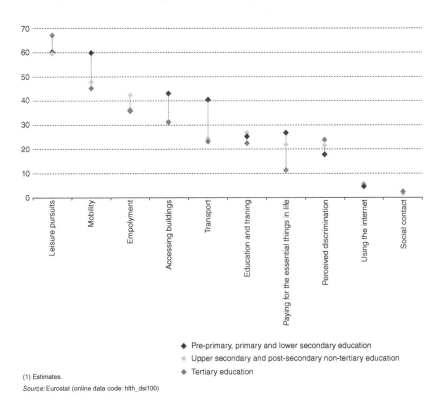

(1) Estimates.

Source: Eurostat (online data code: hlth_dsi100)

Figure 10.3 Share of disabled persons aged 15 and over with a disability in the specified life areas, by educational attainment, EU-27, 2012 (as a percentage of persons reporting a disability in at least one area).

Source: Eurostat 2018.

conducted in Poland (Duplaga 2017) where the author found statistical evidence of the extent to which disabled people lag behind online activities.

Despite the factors determining the use of the Internet amongst disabled people being similar to those of the general population (e.g. place of residence, level of education, occupational status, net income), people with disabilities face a significant digital divide.

Conclusions

Mobility lies at the very heart of people's identities, opportunities and general life experiences. Yet, the research to understand the various perspectives of disability as a concept and social construct is only at the beginning (Goggin 2016). All in all, despite the conclusion that disabled people travel less than non-disabled even if they have similar needs (Bascom 2017, Transport for London 2014) and are granted incentives for using public transport (e.g. the UK Freedom Pass), it seems reasonable to assume by looking at previously mentioned dissimilar trends captured by several studies that there is no general agreement nor clear understanding about most disabled population mobility habits. People with a disability are a very heterogeneous group that may have several different types of impairment which inhibit their travel options differently and consequently their personal quality of life and independence. As such, this group is an indication of how contextual and relational mobility poverty actually is.

Proposing solutions to overcome mobility limitations faced by disabled people is far from trivial. Foremost, mobility is a multi-dimensional, multi-layered phenomenon. Secondly, there are many manifestations of disabilities, impacting a person's mobility capacity differently. Thirdly, the very notion of justice is deeply rooted in the norms and values of societies. Different justice theories lead to different policy measures, based on certain minimum thresholds that are valid for all or measures that enable specific vulnerable groups. This substantially leads to the question of whether accessibility or mobility of different groups is only absolutely, and also relationally, improved with respect to the highly mobiles or hypermobiles.

Other challenges loom ahead. General measures for different types of disability may be considered as a form of reducing the right to individualism or individual freedoms. Such freedoms lay at the core of liberalism and individualism, which are mainstream ideologies in most developed societies. On the other hand, tailored measures (i.e. affirmative actions) aimed at raising the mobility of the disabled are also potentially stigmatising. They will perpetuate and emphasise the inferior status of the disabled, due to the implementation of "special" measures. The creation of special mobility services like ADA paratransit in the US is precisely such an example one could argue because it has the aim of integrating/including disabled people into mainstream society, but, by confining their mobility to these special services, such people are actually prevented from socialising and participating in normal everyday life activities. Ultimately, such good measures are worsening the gap between the disabled and non-disabled people. Moreover, such measures are, in practical terms, determining hierarchies of places where disabled people can and cannot participate. This is raising a new plethora of concerns linked with taking away a person's fundamental and universal right of mobility. Mobility measures concerning disabled people are commonly taken by transportation and planning experts who lack the required competences, authority or sensitivity.

There are no easy answers nor prescriptive measures to overcome the mobility impairment of disabled people. Moreover, as discussed earlier, disabled people often live in ghettoised regions with insufficient levels of participation and involvement in civil society (Imrie 2000). The development of new perspectives and political programmes are thus required to overcome the current injustice and raise the mobility of disabled people to satisfactory levels. Ultimately, disabled people will depend on specific factors such as the social, political and institutional structures of their local geographical contexts. After all, immobility rips a person from her/his fundamental human elements.

Notes

1 https://www.un.org/development/desa/disabilities/convention-on-the-rights-of-persons-with-disabilities.html
2 When the question relates to a person conducting basic activities, it can mean that we are referring to different types of barriers for any given person who reports a disability. It is possible that the same person may, ultimately, experience obstacles or barriers in several different types simultaneously.
3 It is interesting to cite here one of the most important results about the value for money delivered by the concessionary scheme in force in the UK, which altogether came across with a benefit cost ratio of 2.8, which shows to what extent benefits outweigh costs.
4 EU-SILC consists of a multi-purpose instrument which has its focus mainly on income, with detailed data being collected on income components, mostly on personal income. This detailed data collection will also retrieve data information on social exclusion, housing conditions, labour, education and health information.

References

Bascom, Graydon. 2017. "Transportation related challenges for persons' with disabilities social participation." Utah State University. https://digitalcommons.usu.edu/etd/5265. Accessed 9 July 2018.

Bowtell, James. 2015. "Assessing the value and market attractiveness of the accessible tourism industry in Europe. A focus on major travel and leisure companies." In *Journal of Tourism Futures* 1, no. 3: 203–222. https://doi.org/10.1108/JTF-03-2015-0012.

Cuthbertson, Anthony. 2015. Exoskeletons v wheelchairs: Disability advocates clash with futurists over\'offensive\' solution. In *International Business Times*. https://www.ibtimes.co.uk/exoskeletons-vs-wheel-chairs-disability-advocates-clash-futurists-over-offensive-solution-1496178. Accessed 2 April 2020.

DPTAC. 2002. "Attitudes of disabled people to public transport research study conducted for disabled persons transport advisory committee." https://trimis.ec.europa.eu/sites/default/files/project/documents/20060811_110503_45123_UG395_Final_Report.pdf. Accessed 2 April 2020.

Duplaga, Mariusz. 2017. "Digital divide among people with disabilities. Analysis of data from a nationwide study for determinants of Internet use and activities performed online." In *PloS One* 12, no. 6: e0179825–e0179825. https://doi.org/10.1371/journal.pone.0179825.

European Parliament and the Council of the European Union. 2006. "Regulation (EC) No 1107/2006 concerning the rights of disabled persons and persons with reduced mobility when travelling by air." European Parliament; Council of the European Union. https://op.europa.eu/en/publication-detail/-/publication/88a98652-688f-47ff-b79a-e55231b96a2a. Accessed 2 April 2020.

Eurostat. 2015. "Disability statistics – poverty and income inequalities." In *Disability Statistics – Poverty and Income Inequalities*. https://ec.europa.eu/eurostat/statistics-explained/index.php/Disability_statistics_-_poverty_and_income_inequalities. Accessed 2 April 2020.

Eurostat. 2018. "Disability statistics – barriers to social integration – statistics explained." https://ec.europa.eu/eurostat/statistics-explained/index.php?title=Archive:Disability_statistics_-_barriers_to_social_integration. Accessed 2 April 2020.

Gleeson, Brendan. 2006. "Changing practices, changing minds." In *NDA 5th Annual Conference Civil, Cultural and Social Participation: Building and Inclusive Society*, Dublin, Ireland. www.nda.ie. Accessed 2 April 2020.

Goggin, Gerard. 2016. "Disability and mobilities: Evening up social futures." In *Mobilities* 11, no. 4: 533–541. https://doi.org/10.1080/17450101.2016.1211821.

Greener Journeys. 2014. "The costs and benefits of concessionary bus travel for older and disabled people in Britain." https://greenerjourneys.com/wp-content/uploads/2014/09/Concessionary-travel-costs-and-benefits-September-2014.pdf. Accessed on 9 July 2018.

Imrie, Rob. 2000. "Disability and discourses of mobility and movement." In *Environment and Planning A* 32, no. 9: 1641–1656. https://doi.org/10.1068/a331.

Kenyon, Susan, Glenn Lyons, and Jackie Rafferty. 2002. "Transport and social exclusion. Investigating the possibility of promoting inclusion through virtual mobility." In *Journal of Transport Geography* 10, no. 3: 207–219. https://doi.org/10.1016/S0966-6923(02)00012-1.

Lucas, Karen. 2012. "Transport and social exclusion. Where are we now?" In *Transport Policy* 20: 105–113. https://doi.org/10.1016/j.tranpol.2012.01.013.

Oliver, Michael. 1990. *The Politics of Disablement*. London: Palgrave.

Paez, Antonio, Ruben G. Mercado, Steven Farber, Catherine Morency, and Matthew Roorda. 2009. *Mobility and Social Exclusion in Canadian Communities*. Toronto: Policy Research Directorate Strategic Policy and Research.

Preston, John, and Fiona Rajé. 2007. "Accessibility, mobility and transport-related social exclusion." In *Journal of Transport Geography* 15, no. 3: 151–160. https://doi.org/10.1016/j.jtrangeo.2006.05.002.

Pyer, Michelle, and Faith Tucker. 2014. "'With us, we, like, physically can't'. Transport, mobility and the leisure experiences of teenage wheelchair users." In *Mobilities* 12, no. 1: 36–52. https://doi.org/10.1080/17450101.2014.970390.

Rebstock, Markus, ed. 2017. "Economic benefits of improved accessibility to transport systems and the role of transport in fostering tourism for all." International Transport Forum Discussion Paper. https://www.itf-oecd.org/sites/default/files/docs/improved-accessibility-fostering-tourism-for-all.pdf. Accessed 9 July 2018.

Samek Lodovici, Manuela, and Nicoletta Torchio. 2015. "Social inclusion in EU public transport." Policy Department B: Structural and Cohesion Policies – European Parliament. Brussels. http://www.europarl.europa.eu/RegData/etudes/STUD/2015/540351/IPOL_STU(2015)540351_EN.pdf. Accessed 13 June 2018.

Thorslund, Birgitta. 2014. *Effects of Hearing Loss on Traffic Safety and Mobility.* Linköping: Linköping University Electronic Press. http://liu.diva-portal.org/smash/get/diva2:762084/FULLTEXT01.pdf. Accessed 2 April 2020.

Transport for London. 2014. "Understanding the travel needs of London's diverse communities. A summary of existing research." http://content.tfl.gov.uk/understanding-the-travel-needs-of-london-diverse-communities.pdf. Accessed 13 June 2018.

United Nations, Department of Economic and Social Affairs. 2008. "Convention on the rights of persons with disabilities. United Nations." https://www.un.org/development/desa/disabilities/convention-on-the-rights-of-persons-with-disabilities/convention-on-the-rights-of-persons-with-disabilities-2.html. Accessed 6 May 2020.

Urry, John. 2007. *Mobilities.* Cambridge: Polity Press.

11 Migrants, ethnic minorities and mobility poverty

Patrick van Egmond, Tobias Kuttler and Joanne Wirtz

Abstract

Very few studies in Europe have investigated the mobility needs and travel patterns of migrants or have included ethnic perspectives. Apart from generally having less access to cars, barriers in accessing public transport can also be identified. Among these barriers are language barriers, availability and accessibility issues, the cost of public transportation and racial and religious discrimination.

This chapter investigates the existing literature on mobility patterns and problems of migrants in Europe and beyond. In a second step, previous insights are complemented by findings from the authors' own research. The HiReach project investigated the mobility needs and problems of refugees in Luxembourg and Germany. This research provides a rare insight into the mobility poverty experiences of migrants who recently arrived in Europe.

This chapter will outline the mobility challenges first and will then highlight some approaches to solutions in the final part.

Migrants, ethnic minorities and mobility

The travel behaviour of migrants in the environment of their destination country is a neglected area of statistical information and research. There is little knowledge regarding the travel behaviour and attitudes of immigrants in Europe towards different travel modes, due to scarce data and information. Trying to grapple with the research gap, this chapter will provide an insight into the number of migrants and ethnic minorities presently residing in the EU, their socio-economic situation, the specific needs and socio-cultural characteristics of migrants and ethnic minorities in relation to mobility and transport. To further close the research gap, links will be drawn from existing literature to the results from two case studies in Germany and Luxembourg in the HiReach project.[1] It will show how a mismatch between these needs and the present mobility offer might lead to an increased risk of exclusion. A number of solutions are discussed, such as fare reduction and familiarisation with alternative forms of mobility, such as cycling. Finally, we will look at changing perceptions and transportation usage among migrant and ethnic minorities in order to estimate the impact of digitisation and the debate on climate change.

Migrants and ethnic minorities in the EU

Migration is presently much debated in the European Union. This specifically focuses on migration from the Middle East and the African continent. For the EU, a migrant person is "a broader-term of an immigrant and emigrant that refers to a person who leaves from one country or region to settle in another, often in search of a better life" (European Commission 2016). The EU's official position towards migrants is that they represent an important part of the development of European societies, both in economic, social and cultural perspectives. Meanwhile, there is fear from part of the population about the effects of migration, specifically from non-EU countries, on their society and living conditions.

In 2017, 2.4 million immigrants entered the EU from non-EU countries. This brought the number of non-EU citizens up to 22.3 million persons (4.4% of the 512.4 million people living in the EU). The EU Member States granted citizenship to 825 thousand persons in 2017 (Figure 11.1).

Securing successful acclimatisation for migrants at their places of destination is for the benefit of both migrant and domestic societies, and maximises the positive effects of legal immigration to EU development. It is nevertheless clear that in many situations migrants are vulnerable populations, due to a myriad of possible reasons.

Some of the migrants can also be included in the definition of ethnic minority, even if most of the ethnic minorities in most European Member States have been part of societies for more than a generation, sometimes even for several generations.

Ethnic minorities represent a group for whom social and economic exclusion remain an everyday challenge in Europe today. Cultural and ethnic

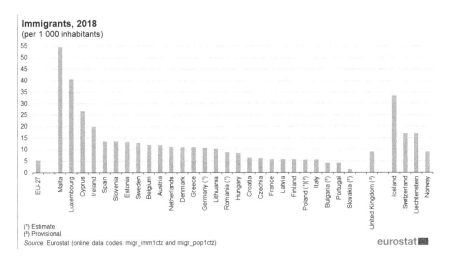

Figure 11.1 Share of Immigrants in EU Member States, 2018.
Source: Eurostat 2019.

backgrounds shape the challenging differences between ethnic minorities and majority populations. These differences are often also reflected on the labour market (unemployment, underemployment and substandard remuneration), uneven access to health care and social services. Especially relevant are the connections between appropriate quality of life, prosperity and social cohesion and the absence of significant labour market segmentation (Kahanec et al. 2010).

Figure 11.2 shows that in many countries Europeans represented almost half of all the foreign-born people who lived in an EU Member State. Some European cities account for very large proportions of migrant or ethnic minority populations. For example, London has an estimated 40% "Black, Asian and minority ethnic (BAME)" population (Transport for London 2014), while Berlin has an estimated 18% non-European population. Interestingly, while some regions attract most of their migrants from a narrow range of countries, others are extremely diverse, drawing migrants from around the world. This is particularly true for some of Europe's largest cities and capital cities, for example, the regions of Hamburg, Munich, Paris, Amsterdam, Stockholm and London. Geographic proximity, ex-colonial links, common languages and cultural ties play an important role in determining the destinations favoured by migrants (Tsang and Rohr 2011).

Analysing recent Eurostat statistics regarding migrants, it is possible to characterise them properly. For example, men represented 54% of the immigrants to the EU Member States in 2018. This share was the highest in Croatia (75%) and the lowest in Portugal (53%) (Eurostat 2020).

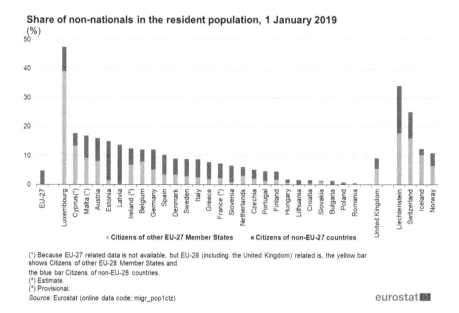

Share of non-nationals in the resident population, 1 January 2019

(¹) Because EU-27 related data is not available, but EU-28 (including the United Kingdom) related is, the yellow bar shows Citizens of other EU-28 Member States and the blue bar Citizens of non-EU-28 countries.
(²) Estimate.
(³) Provisional.
Source: Eurostat (online data code: migr_pop1ctz)

eurostat

Figure 11.2 Share of non-nationals in the resident population, 1 January 2019.
Source: Eurostat 2019.

Table 11.1 Foreign-born population by country of birth, 1 January 2017

	Total		Citizens of another EU-27 Member State		Citizens of a non-EU-27 country		Stateless	
	(Thousand)	(% Of the population)	(Thousand)	(% Of the population)	(Thousand)	(% Of the population)	(Thousand)	(% Of the population)
Belgium	1,400.2	12.2	900.6	7.9	498.6	4.4	1.0	0.0
Bulgaria	95.8	1.4	9.6	0.1	84.3	1.2	1.9	0.0
Czechia	557.5	5.2	225.4	2.1	332.1	3.1	0.0	0.0
Denmark	525.8	9.1	205.7	3.5	311.8	5.4	8.3	0.1
Germany	10,089.3	12.2	4,293.9	5.2	5,783.9	7.0	11.5	0.0
Estonia	199.2	15.0	20.1	1.5	179.1	13.5	0.0	0.0
Ireland[a]	612.0	12.5	336.7	6.9	275.0	5.6	0.3	0.0
Greece	831.7	7.8	196.7	1.8	635.0	5.9	0.0	0.0
Spain	4,840.2	10.3	1,679.7	3.6	3,158.7	6.7	1.6	0.0
France[b]	4,882.6	7.3	1,460.9	2.2	3,421.7	5.1	0.0	0.0
Croatia	66.5	1.6	17.2	0.4	48.5	1.2	0.8	0.0
Italy	5,255.5	8.7	1,554.0	2.6	3,700.7	6.1	0.8	0.0
Cyprus	155.6	17.8	:	:	:	:	0.0	0.0
Latvia	266.6	13.9	6.1	0.3	260.3	13.6	0.2	0.0
Lithuania	47.2	1.7	6.9	0.2	39.2	1.4	1.0	0.0
Luxembourg	291.3	47.4	240.3	39.1	50.8	8.3	0.2	0.0
Hungary	180.5	1.8	71.2	0.7	109.3	1.1	0.1	0.0
Malta	83.3	16.9	:	:	:	:	0.0	0.0
Netherlands	1,068.1	6.2	520.4	3.0	534.8	3.1	12.9	0.1
Austria	1,427.1	16.1	719.2	8.1	703.4	7.9	4.4	0.0
Poland	2,898	0.8	29.0	0.1	260.2	0.7	0.6	0.0
Portugal	480.3	4.7	132.5	1.3	347.8	3.4	0.0	0.0
Romania	121.1	0.6	57.8	0.3	63.0	0.3	0.3	0.0
Slovenia	138.2	6.6	20.1	1.0	118.1	5.7	0.0	0.0
Slovakia	76.1	1.4	56.1	1.0	18.5	0.3	1.5	0.0
Finland	256.0	4.6	95.1	1.7	159.7	2.9	1.2	0.0
Sweden	920.1	9.0	302.0	3.0	598.4	5.8	19.8	0.2
United Kingdom	**6,171.9**	**9.3**	**3,681.9**	**5.5**	**2,490.1**	**3.7**	**0.0**	**0.0**
Iceland	44.3	12.4	36.5	10.2	7.7	2.2	0.0	0.0
Liechtenstein	13.1	34.0	6.9	17.9	6.2	16.1	0.0	0.0
Norway	584.1	11.0	343.6	6.4	237.8	4.5	2.6	0.0
Switzerland	2,146.4	25.1	1,370.4	16.0	775.5	9.1	0.5	0.0

Source: Eurostat (2018d).

Note: The values for the different categories of citizenship may not sum to the total due to rounding. Cyprus and Malta are not displayed because no detailed data by individual country are available

a Estimate.
b Provisional.

Regarding the age of immigrants, they were considerably younger (with a median age of 27.9 years in 2011) than the total population already residing in their country of destination (with a median age of 42.9 years) (Eurostat 2016) (Figure 11.3).

Being part of a migrant or minority ethnic group is likely to increase the risks of marginalisation and poverty, namely due to factors such as discrimination, racism and cultural and language problems. These (especially if combined) reduce access to good-quality jobs and education and increase the likelihood of living in deprived areas. Within these groups, women are particularly at risk.

Data from an EU-28 survey in 2017 shows that one in four (24%) respondents felt discriminated against because of their ethnic or immigrant background in the 12 months preceding the survey. The highest rates of discrimination based on ethnic or immigrant background are observed in the area of employment and when accessing public and private services (European Union Agency for Fundamental Rights 2017), namely health, education and transportation.

Mobility behaviour and mobility-related disadvantages of migrants and ethnic minorities

As mentioned earlier, very few studies in Europe have investigated the mobility needs and travel patterns of migrants or have included ethnic perspectives in their studies. Therefore, it is necessary to expand the scope and also include studies from other parts of the world.

Comprehensive studies on the relationships between travel behaviour and immigrant status focusing on the distinctive travel patterns of immigrants have been conducted in the United States (see e.g. Myers 1997; Rosenbloom and Fielding 1998; Deakin et al. 2002; Purvis 2003; Casas et al. 2004; Handy and Tal 2005; Chatman and Klein 2009). These studies suggest that in the first five to ten years of living in the United States, the travel behaviour of immigrants is different from the behaviour of citizens born in the United States, but immigrants often assimilate after just five years. Also, usage of public transportation differs among immigrant groups, regardless of the amount of time they had been residing in the United States. Although there is usually a preference for private modes of transport such as cars, (Blumenberg 2008; Lovejoy and Handy 2008), some groups, e.g. Latino communities, are more open to using collective modes of transport (Douma 2004; Valenzuela et al. 2005). This includes car-pooling, sharing rides and borrowing cars in social networks, as identified by Lovejoy and Handy (2011) in the case of Mexican immigrants in California. Tal and Handy (2010) suggest that mode preferences stem partially from attitudes based on previous experiences in the country of origin and that different travel behaviours are also a result of the social segregation of immigrants. The spatial location of residence and residential choices also have an impact on commuting behaviour, as Beckman and Goulias (2008) point out for the case of California.

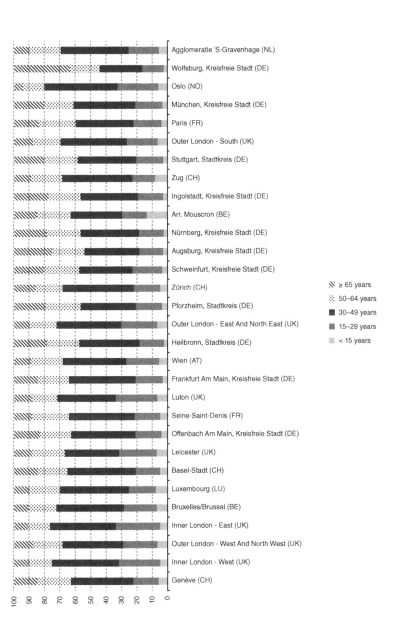

≥ 65 years
50–64 years
30–49 years
15–29 years
< 15 years

(¹) The figure shows the top 30 regions in the EU-28, Norway and Switzerland with the highest proportions of foreign-born inhabitants, based on NUTS 2010 level 3 regions that are either predominantly urban regions or which correspond to (part of) a metropolitan region. The regions are ranked descending on the share of foreign-born inhabitants in the total regional population.

Figure 11.3 Distribution of foreign-born inhabitants by age, selected urban regions, 2011 (% of foreign-born inhabitants).

Source: Eurostat 2016.

For two metropolitan areas in Australia, Klocker et al. (2015) observed below-average rates of car ownership and use amongst ethnic minorities and migrants. However, they found that these are not the result of socio-economic deprivation, but rather a result of cultural factors and experiences of transport in countries of origin as well as the choices for residential location.

Insights into travel behaviour and mode choice of different ethnic minorities and migrant groups in Europe are largely absent. For the United Kingdom, Rajé (2004), referring to other older small-scale studies, summarised attitudes towards the different modes of transport of women in London. While all women felt unsafe walking at night, they would usually feel safe during the day. The transport usage of women differed substantially between women of Asian origin and those of Afro-Caribbean origin.

Several studies in Europe focussed on accessibility to public transport for migrant groups and ethnic minorities. These studies highlight that the availability of public transport at affordable fares has effects on employment opportunities and access to basic services. This is particularly relevant considering that migrant populations are more prone to low incomes and unemployment (Samek Lodovici and Torchio 2015). Apart from being more at risk of ill health, poverty and unemployment, ethnic minorities are more likely to live in deprived areas as well as in overcrowded or unpopular housing (Rajé 2017).

For groups at risk of social exclusion, remoteness can also have a significant impact on their quality of life, as they often do not have access to cars.

In relation to the use of public transport, the following barriers were identified in the "Together on the Move" project (Assum et al. 2011):[2]

- Language barriers;
- Availability and accessibility issues;
- Costs; and
- Racial and religious discrimination.

It has to be stated that, beyond low incomes and other barriers, there are also matters of racial and religious discrimination that have an effect on personal mobility and make the usage of (public) transportation an unfortunate experience. It was already found in earlier research that even migrants who were actually originally from middle to upper class environments in their country of birth were nevertheless experiencing social exclusion in their migration destination. Migrants of South Asian origin living in London reported that they are facing discrimination not only at their workplace but also while travelling on public transport. Living at the margins of society, they almost never leave the areas around their houses (Rutten and Verstappen 2014). Rajé (2004, 40) also reports from a study about the perception of crime and safety in London that black and minority ethnic communities exhibit higher levels of fear than non-minority groups, partially due to the higher levels of harassment they experience.

Other forms of discrimination were reported from all over the United Kingdom, such as bullying of black school children, bus drivers not stopping for members of ethnic communities, attacks on black bus drivers, to name a few frequent incidents (Beuret et al. 2000, cited in Rajé 2004, 42–43). Negative perceptions of public transport can be partially traced back to experiences in the country of origin, as already highlighted above; Rajé (2004, 47), for instance, reports about negative attitudes of migrants from the West Indies in the United Kingdom, due to the dangerous circumstances and unreliability of public transport in Jamaica and other parts of the Caribbean.

The HiReach research also found that – in the case of Luxembourg – young immigrants or ethnic minorities and immigrant women wearing headscarves often say that they are the first ones to be checked at inspections.

Given that these groups are highly dependent on public transport, but have difficulties in accessing those services that cater to their needs, it was brought up in the focus group discussion that there is the probability of switching to private second-hand vehicles at some point. However, these exhibit increased emissions and air pollution.

A particularly vulnerable group among ethnic minorities and migrant groups are the elderly. In the United Kingdom, access to cars is lower among the elderly, especially among communities of Asian origin, but they also have negative perceptions of public transport. Many elderly from ethnic minorities in the United Kingdom were unaware that they are entitled to travel allowance (Rajé 2004, 40–41).

Overall, the travel patterns of ethnic minorities and migrants are still poorly understood in Europe. In research projects, members of such groups are often hard to reach. Also, when there are consultations for specific transport development projects, members of ethnic communities may not participate because appropriate channels of communication between authorities and the communities are not in place and hence the members may not have heard about these plans (Rajé 2004, 81–82, 94–97, 2007, 67). Rajé also highlights the "experience gap" between the experiences of transport system users and the understandings of users' experiences held by planners and policy-makers. Besides gender bias, this can in particular be traced back to the ethnic composition of the transport planning profession (Rajé 2007, 52).

Fiona Rajé has criticised a lack of cultural awareness among transport professionals and has therefore called for a micro-understanding of the role of ethnicity in mobility patterns. Taking such perspectives into account, she argues that "complexities around journeying at every stage of the process" should be discussed in transport policy literature (Rajé 2004, 47). She highlights that when not aware of cultural specificities, the practices of social seclusions especially of women would not be taken into account. This has the effect that some dimensions between ethnicity, transport and social exclusion would remain hidden, for example that women rely on men or women of different ethnic groups to fulfil certain everyday tasks (Rajé 2004, 49–51, 64–67).

Mobility needs and the problems of refugees in Luxembourg and Germany

During the course of the HiReach research, it became evident that migrants and ethnic minorities should not be treated as a homogeneous group. Many are well integrated in society as the result of having already been in their host country for a few years, different economic situations as well as their ability to adapt more quickly to the mobility and transport habits in the country. Mobility poverty among migrants and ethnic groups combines the difficulties of those newly arriving in a country with other aspects of vulnerability (e.g. old age, disability, low income).

The HiReach project investigated the particular mobility needs and problems of **refugees** in two regions of Luxembourg and Germany. The HiReach project therefore provides a rare insight into the experiences of mobility poverty of migrants who recently arrived in Europe. Besides material deprivation and language barriers, this study also took the living, employment and education situation of refugees into account. Specific attention was given to the mobility situation of women, since earlier work shows that having good access to public transport is very relevant for women (Uteng 2009).

HiReach research shows that newly arriving refugees are heavily reliant on public transport, which is costly for them if they have to pay for it themselves. For those having direct and convenient access to public transport, it is easier to find opportunities for employment, education and meeting friends.

The place of accommodation is in this regard essential and can have a significant effect on everyday mobility. Initial accommodation for refugees – that is provided by the state – is considered as a "lottery" from the viewpoint of the refugees. Such accommodation may be assigned in urban centres or in rural areas where there is less public transport connectivity. The personal situation and characteristics are usually not considered when accommodation is assigned to them and this potentially hinders them in their everyday life. When refugees need to find accommodation on the conventional housing market, they often face difficulties. In many urban areas in Germany, housing is so expensive that refugees are forced to live in areas less well served by public transport.

HiReach research suggests that refugees would use public transport more often, and would travel more overall, if they had a price-reduced monthly ticket. Especially for newcomers, a free ticket would significantly reduce their travel budget and, since expenses for mobility eat up a large part of their monthly budget, it would also ease their financial situation overall. Some cities in Germany have recognised this need, for example Berlin, where the "Welcome to Berlin" ticket is offered at a heavily subsidised rate. Luxembourg has made its whole public transport free of charge for all users.

However, refugees participating in the research indicated that they would be willing to pay for public transport if that would increase its availability and more efficiently meet their mobility needs. This confirms that mobility is only partly limited by financial constraints, especially when it comes to leisure trips. Interestingly, half of those who participated in the HiReach research stated that they would not walk or use their bikes less if they had a cheaper public transport ticket.

Earlier research shows similar challenges among non-EU migrants, independently of their year of arrival, socio-demographic characteristics or place of residence and work (Tsang and Rohr 2011).

A free or public transport fare at a reduced rate should be intelligently combined with good access to transport. For example, the HiReach focus groups discussed the case of the Stuttgart social ticket. This ticket offers a reduction of half the ticket price for monthly tickets for the inner zones, but, for monthly tickets over larger distances, the reduction is less than a third. Hence, in such a pricing structure, those who have to travel large distances for work or education, which is often the case for migrants and ethnic minorities, are more disadvantaged than those located in an urban area.

Nevertheless, it is clear and was also voiced that a social ticket is better than no fare reduction at all. It was confirmed that a monthly ticket at a reduced price would enable them to find a job or place of education more easily.

It became clear that for education (school, study, training) migrants and ethnic minorities have to travel long distances, but shopping, activities with children and other purposes are usually conducted within walking distance.

In relation to the purpose of travel, there were specific findings on the mobility problems of women refugees. Women have to travel longer distances than men, e.g. to attend a specialist doctor. The language problem seems to be more severe for women because some of them are illiterate and have difficulties in using a ticket machine. Women also travel more often with young children on public transport.

In general, for recently arrived refugees, it takes some time to understand the complex tariff and ticket system on public transport even if there are often local volunteers and the social service for refugees to support newcomers by explaining how the public transportation system works.

One finding not to be neglected was that refugees from African countries reported that they face discrimination on public transport. It seems that, especially when young African men are travelling in groups, they are suspicious to ticket inspectors and checked more often.

Riding a bicycle in particular, as with public transport, appears to be regarded as an inferior form of transport at least by certain immigrant groups. Cycling appears to be more appealing to locally born people than to immigrants, especially women (Samek Lodovici and Torchio 2015). Cycling is common among male refugees. In particular situations, e.g. when public transportation is not available on weekends or during the night/early/late hours, cycling is also used to travel between urban areas. Even though over half of refugee women know how to ride a bicycle, bike usage among these women is not common. The remaining women expressed that they would be eager to learn how to ride a bicycle.

Several newly arriving citizens have a driving licence or are undergoing driver training and exams. Nevertheless, car ownership is lower among immigrants, which can be suggested to be related to their less favourable economic conditions (Samek Lodovici and Torchio 2015) and they travel less in general. Trips are fewer and travel distances by car are shorter among immigrants than among the domestically born populations (Assum et al. 2011). Licence holding, car ownership and car use are not strictly related, namely because deciding to buy a car or to get a driving licence derives from individuals' needs, financial constraints and preferences (Tsang and Rohr 2011).

The differences in public transport access, cycling and car usage between migrants and those domestically born seem to be greater for women than for men and greater for newly arrived migrants than for migrants who have stayed longer in their new country. This confirms earlier work on this topic (Assum et al. 2011; Samek Lodovici and Torchio 2015).

It shows that pre-existing attitudes towards different modes combined with improved economic standards among immigrants over time leads to higher car access among immigrants and consequently to less sustainable travel (Assum et al. 2011). The HiReach work showed that, as a result of the present climate concerns and debate, this position is evolving. Climate change is not neglected within the groups of refugees and ethnic minorities. Specifically, the younger migrants in Luxembourg also expressed environmental concerns as a reason to use more sustainable modes of transport.

Specifically, cycling is seen more and more as a possible solution to their transport needs. Some refugees involved in the HiReach research did not know how to cycle, especially the elderly and women. On the other hand, access to a bicycle is usually available (i.e. their own bicycle or a public bicycle). It was pointed out that many newcomers are eager to learn how to ride a bicycle and that it is important to learn how to cycle in car traffic.

A cycling training programme such as Fietsmeesters in the Netherlands trains people how to ride a bicycle. It provides easier accessibility to transportation because of the minimal cost of cycling compared to other transport modes. Also, the maintenance cost of a bicycle is considered quite low. In Germany, the mobility of refugees is supported by local volunteer organisations collecting used bicycles among the residents of towns and villages. Members of the volunteer groups repair the bicycles and donate them to newly arriving refugees. Furthermore, some of these volunteer groups provide bicycle repair and maintenance facilities for refugees and train refugees to repair their bikes themselves (Figure 11.4).

As mentioned above, a number of participating women knew how to cycle. A Syrian participant mentioned that in Syria a lot of women know how to ride a bicycle, especially younger women from urban areas. A cycling programme for elderly people would be particularly interesting as a fun physical leisure activity (Figures 11.5 and 11.6).

(a)   (b)

Figure 11.4 Die Fahrradfüchse ("bicycle foxes") is a volunteer organisation in the town of Donzdorf in southern Germany. With support from the municipality, it offers bicycle repair and maintenance facilities for refugees. These facilities are also open to other low-income persons in the region.

Source: Tobias Kuttler, TU Berlin.

Figure 11.5 Focus group sessions with refugees in Esslingen (Neckar)/Germany.

Sources: Tobias Kuttler, TU Berlin.

Figure 11.6 Focus group session with migrants and refugees in Luxembourg.
Sources: Patrick van Egmond/LuxMobility, Luxembourg.

Openness towards ICT-supported mobility solutions among refugees

HiReach enabled an understanding of the effects of mobilities and related perceptions within groups of refugees and migrants to be provided as a result of the introduction of ICT and new forms of mobility.

The internet is a major source of information especially for newly arriving migrants and refugees. They are often frequent users of ICT. Furthermore, social media is vital to maintain connections with families and friends at their places of origin (see e.g. Charmakeh 2013; Harney 2013). Studies in Australia and New Zealand found that newly arriving refugees have largely positive perceptions of ICT, its usefulness and importance; they use ICT extensively for finding accommodation and jobs, being involved in new social connections and networks, and – finally – organising their daily mobility (Kabbar and Crump 2006; Felton 2015). However, the study also found that there are significant language, literacy and cultural barriers for many refugees when using ICT solutions (Alam and Imran 2015; Felton 2015).

ICT is particularly important for refugees in organising travel and navigating the city (Felton 2015). In a study about Syrian refugees in Germany, it was highlighted that smartphone usage helped refugees to overcome language barriers, especially when travelling in the city. Being able to travel, with the support of ICT, contributed to a sense of agency and well-being among refugees, thereby also enhancing perceptions of social inclusion (AbuJarour and Krasnova 2017).

In our own research, it was found that migrants had more mobility needs in terms of travel to the city centre for education and administration. Mobility needs also included seeing friends, personal appointments, education and sport. It showed that the newly arriving migrants seem to have fewer issues in fulfilling those needs as they were able to better understand the public transport network in comparison to the elderly resident population and older migrants. Younger migrants in particular are well acquainted with internet and smartphone applications and it is easy for them to find fares and information on public transport services including new forms of mobility (e.g. public bicycles). However, the HiReach work points out that newcomers are sometimes confused about the rules regarding tickets and that they had consequently been in trouble with ticket inspectors.

Also, all types of ride-sharing solutions seem to have a potential to reduce mobility poverty among migrants and ethnic minorities to a certain extent. First, as described earlier, due to the high perceived costs of travelling on public transportation, ride-sharing can be a cost-effective alternative. Second, many refugees need to travel at odd times (late evening, early morning, weekends) for their jobs or want to visit friends in other parts of the region where public transport connections are less frequent or cumbersome with many interchanges. Third, it became clear that refugees intensively use ride-sharing for long-distance trips, such as BlaBlaCar, hence they could easily relate to local ride-sharing solutions for local and regional travel. Finally, especially the younger migrants make intensive use of smartphones so that a smartphone-based ride-share solution would be less of an issue for them.

In the focus group sessions with refugees, a ride-sharing solution called "Fairfahrt" was discussed. Fairfahrt (fair-ride in German) is a ride-sharing platform organised in and around the small town of Romrod, located in the rural area of Hesse, Germany.[3]

Generally, the migrants participating in the focus group stated that local ride-sharing is a good idea and that it would be a way to use ride-sharing in the city and region. Most of the refugees already use BlaBlaCar, but services on this platform are usually not for short distance travel. A system like Fairfahrt would supplement public transport at times of low frequency or places with low public transport coverage.

Younger migrants in particular are all well experienced in using their smartphones for organising travel. Hence, they felt comfortable in understanding and using ride-sharing services. This was supported by statements that they usually enjoy ride-sharing and interacting with other people while travelling. The participants highly appreciated that one of the explicit aims of the Fairfahrt system is to bring people together at local/regional level who would otherwise not meet.

As a system like Fairfahrt is free for riders, the service is attractive to refugees who have financial constraints. Hence, they highly value ride-sharing solutions for their cost effectiveness. Nevertheless, it was felt that ride-sharing should be seen as a complement of public transport. Being on time for appointments at the authorities, at the doctor's, at school or at work is of utmost importance, due to the constant (perceived or real) threat of losing jobs or being reprimanded in any way. Since it is not guaranteed that a trip request will be answered by a car driver, or may be answered too late, it was argued that it is not reliable enough for certain trip purposes. Finally, as one migrant argued, since the service is on a voluntary basis, drivers may reject a rider, which is different to public transport services that are obliged to transport people (given a valid ticket is purchased). This could lead to uncomfortable situations, including acts of racism. However, the refugees highlighted that they had only good experiences with ride-sharing over long distances, such as BlaBlaCar.

Notes

1 These findings presented here are additional to – but partially overlap with – the findings presented in the fieldwork part of this publication, more precisely Chapter 18.
2 The "Together on the move" project focussed on immigrants living in Austria, Belgium and Norway.
3 The central idea of Fairfahrt is that car drivers, on their way to a specific place in Romrod or nearby, can pick up persons at one of five stations and offer them a free ride in their own car. Participating users have to register themselves and receive an ID card allowing them to add a riding request at one of the five stations. The main station is inside a supermarket, the other four are in other districts of the administrative area of Romrod. The drivers, who only need to download the app (and do not necessarily need to register themselves), either receive a push notification or are informed by a green lamp above one of the stations that a ride request has been entered into the system.

References

AbuJarour, Safa'a, and Hanna Krasnova. 2017. "Understanding the role of ICTS in promoting social inclusion: The case of Syrian refugees in Germany." In *Proceedings of the 25th European Conference on Information Systems (ECIS), Guimarães, Portugal, 5–10 June 2017*, 1792–1806. http://aisel.aisnet.org/ecis2017_rp/115. Accessed 2 April 2020.

Alam, Khorshed, and Sophie Imran. 2015. "The digital divide and social inclusion among refugee migrants." In *Information Technology & People* 28, no. 2: 344–365. https://doi.org/10.1108/ITP-04-2014-0083.

Assum, Terje, Tina Panian, Paul Pfaffenbichler, Jan Christiaens, Susanne Nordbakke, Haval Davoody, and Sarah Wixey. 2011. "Immigrants in Europe, their travel behaviour and possibilities for energy efficient travel." http://www.together-eu.org/docs/file/together_d2.1_state-of-the-art.pdf. Accessed 4 February 2020.

Beckman, Jarad D., and Konstadinos G. Goulias. 2008. "Immigration, residential location, car ownership, and commuting behavior: A multivariate latent class analysis from California." In *Transportation* 35, no. 5: 655–671. http://dx.doi.org/10.1007/s11116-008-9172-x.

Beuret, K., Aslam, H., Gross, A., Osman, A., and F. Khan. 2000. *Ethnic Minorities and Visible Religious Minorities: Their Transport Requirements and the Provision of Public Transport*. London: Department for Environment, Transport and the Regions.

Blumenberg, Evelyn. 2008. "Immigrants and transport barriers to employment: The case of Southeast Asian welfare recipients in California." In *Transport Policy* 15, no. 1: 33–42. https://doi.org/10.1016/j.tranpol.2007.10.008.

Casas, Jesse, Carlos Arce, and Christopher Frye. 2004. "Latino immigration and its impact on future travel behavior." In *National Household Travel Survey Conference: Understanding Our Nation's Travel.* Washington, DC: Transportation Research Board. http://onlinepubs.trb.org/onlinepubs/archive/conferences/nhts/Casas.pdf. Accessed 4 February 2020.

Charmarkeh, Houssein. 2013. "Social media usage, Tahriib (migration), and settlement among Somali refugees in France." In *Refuge: Canada's Journal on Refugees* 29, no. 1: 43–52. https://refuge.journals.yorku.ca/index.php/refuge/article/view/37505. Accessed 2 April 2020.

Chatman, Daniel G., and Nicholas Klein. 2009. "Immigrants and travel demand in the United States: Implications for transportation policy and future research." In *Public Works Management & Policy* 13, no. 4: 312–327. https://doi.org/10.1177/1087724X09334633.

Deakin, Elizabeth, Ferrell, Christopher, Mason, Jonathan, and John Thomas. 2002. "Policies and practices for cost-effective transit investments: Recent experiences in the United States." In *Transportation Research Record* 1799, no. 1: 1–9. https://doi.org/10.3141/1799-01.

Douma, Frank. 2004. "Using ITS to better serve diverse populations." https://conservancy.umn.edu/bitstream/handle/11299/1138/200442.pdf?sequence=1. Accessed 4 February 2020.

European Commission. 2016. *"EU Immigration Portal Glossary."* https://ec.europa.eu/immigration/glossary_en. Accessed 18 September 2020.

European Union Agency for Fundamental Rights. 2017. "Fundamental Rights Report 2017." https://fra.europa.eu/sites/default/files/fra_uploads/fra-2017-fundamental-rights-report-2017_en.pdf. Accessed 18 September 2020.

Eurostat. 2016. "Urban Europe – statistics on cities, towns and suburbs – foreign-born persons living in cities." https://ec.europa.eu/eurostat/statistics-explained/index.php?title=Urban_Europe_%E2%80%94_statistics_on_cities,_towns_and_suburbs_%E2%80%94_foreign-born_persons_living_in_cities#Foreign-born_populations_in_predominantly_urban_and_metropolitan_regions. Accessed 10 June 2020.

Eurostat. 2019. "Migration and migrant population statistics." https://ec.europa.eu/eurostat/statistics-explained/index.php?title=Migration_and_migrant_population_statistics#Migrant_population:_21.8_million_non-EU-27_citizens_living_in_the_EU-27_on_1_January_2019. Accessed 10 June 2020.

Eurostat. 2020. "Migration and migrant population statistics – Statistics Explained." https://ec.europa.eu/eurostat/statistics-explained/pdfscache/1275.pdf. Accessed 18 September 2020.

Felton, Emma. 2015. "Migrants, refugees and mobility: How useful are information communication technologies in the first phase of resettlement?." In *Journal of Technologies in Society* 11, no. 1: 1–13. https://eprints.qut.edu.au/77888/. Accessed 18 September 2020.

Handy, Susan, and Gil Tal. 2005. "The travel behavior of immigrants and race/ethnicity groups: An analysis of the 2001 national household transportation survey." https://escholarship.org/content/qt4b8382vh/qt4b8382vh.pdf. Accessed 4 February 2020.

Harney, Nicholas. 2013. "Precarity, affect and problem solving with mobile phones by asylum seekers, refugees and migrants in Naples, Italy." In *Journal of Refugee Studies* 26, no. 4: 541–557. https://doi.org/10.1093/jrs/fet017.

Kabbar, Eltahir. F., and Barbara J. Crump. 2006. "The factors that influence adoption of ICTs by recent refugee immigrants to New Zealand". In *Informing Science* 9: 111–121. https://doi.org/10.28945/475.

Kahanec, Martin, Anzelika Zaiceva, and Klaus F. Zimmermann. 2010. "Ethnic minorities in the European Union. An overview." DIW Berlin, German Institute for Economic Research. http://ftp.iza.org/dp5397.pdf. Accessed 2 April 2020.

Klocker, Natascha, Stephanie Toole, Alexander Tindale, and Sophie-May Kerr. 2015. "Ethnically diverse transport behaviours: An Australian perspective." In *Geographical Research* 53, no. 4: 393–405. https://doi.org/10.1111/1745-5871.12118.

Lovejoy, Kristin, and Susan Handy. 2008. "A case for measuring individuals' access to private-vehicle travel as a matter of degrees: Lessons from focus groups with Mexican immigrants in California." In *Transportation* 35, no. 5: 601–612. https://doi.org/10.1007/s11116-008-9169-5.

Lovejoy, Kristin, and Susan Handy. 2011. "Social networks as a source of private-vehicle transportation: The practice of getting rides and borrowing vehicles among Mexican immigrants in California." In *Transportation Research Part A: Policy and Practice* 45, no. 4: 248–257. https://doi.org/10.1016/j.tra.2011.01.007.

Myers, Dowell. 1997. "Changes over time in transportation mode for journey to work: Effects of aging and immigration". In *Decennial Census Data for Transportation Planning: Case Studies* 13: 84.

Purvis, Chuck. 2003. "Commuting patterns of immigrants." CTPP 2000 Status Report, Federal Highway Administration, Bureau of Transportation Statistics, Federal Transit Administration. US Department of Transportation, Washington, DC. http://www.trbcensus.com/newsltr/sr0803.pdf. Accessed 4 February 2020.

Rajé, Fiona. 2007. "The lived experience of transport structure: An exploration of transport's role in people's lives." In *Mobilities* 2, no. 1: 51–74. https://doi.org/10.1080/17450100601106260.

Rajé, Fiona. 2017. *Negotiating the Transport System: User Contexts, Experiences and Needs*. London and New York: Routledge.

Rajé, Fiona, Margaret Grieco, Julian Hine, and John Preston. 2004. *Transport, Demand Management and Social Inclusion: The Need for Ethnic Perspectives.* Aldershot: Ashgate.

Rosenbloom, Sandra, and G. J. Fielding. 1998. "TCRP Report 28: Transit Markets of the Future: The Challenge of Change." *Transit Cooperative Research Program, TRB, National Research Council, Washington, DC.* http://onlinepubs.trb.org/onlinepubs/tcrp/tcrp_rpt_28-a.pdf. Accessed 18 September 2020.

Rutten, Mario, and Sanderien Verstappen. 2014. "Middling migration. Contradictory mobility experiences of Indian youth in London." In *Journal of Ethnic and Migration Studies* 40, no. 8: 1217–1235. https://doi.org/10.1080/1369183X.2013.830884.

Samek Lodovici, Manuela, and Nicoletta Torchio. 2015. "Social inclusion in EU public transport." Policy Department B: Structural and Cohesion Policies – European Parliament. Brussels. http://www.europarl.europa.eu/RegData/etudes/STUD/2015/540351/IPOL_STU(2015)540351_EN.pdf. Accessed 10 February 2020.

Tal, Gil, and Susan Handy. 2010. "Travel behavior of immigrants: An analysis of the 2001 National Household Transportation Survey." In *Transport Policy* 17, no. 2: 85–93. https://doi.org/10.1016/j.tranpol.2009.11.003.

Transport for London. 2014. *Understanding the Travel Needs of London's Diverse Communities. A Summary of Existing Research.* http://content.tfl.gov.uk/understanding-the-travel-needs-of-london-diverse-communities.pdf. Accessed 13 June 2018.

Tsang, Flavia, and Charlene Rohr. 2011. *The Impact of Migration on Transport and Congestion.* Cambridge: RAND Europe.

Uteng, Tanu Priya. 2009. "Gender, ethnicity, and constrained mobility. Insights into the resultant social exclusion." In *Environment and Planning A* 41, no. 5: 1055–1071. https://doi.org/10.1068/a40254.

Valenzuela Jr, Abel, Lisa Schweitzer, and Adriele Robles. 2005. "Camionetas: Informal travel among immigrants." In *Transportation Research Part A: Policy and Practice* 39 no. 10: 895–911. https://doi.org/10.1016/j.tra.2005.02.026.

12 Children and young people

*Stefano Borgato, Silvia Maffii
and Simone Bosetti*

Abstract

Understanding mobility patterns and transportation challenges for children and young people is fundamental in order to give them proper access to education, friends, social activities and set the basis for a bright future. However, juvenile travel needs, especially those non-school related but intended for leisure and recreational activities, are currently not properly addressed. At the same time, the right to have a safe travel environment is often overlooked. This chapter reflects on children's mobility behaviour and goes deeper into the struggles they face to access transport. It also provides a series of considerations on how to make the transport sector more friendly for younger people and to eventually reduce their dependence on their parents.

Introduction

It is fundamental to consider children and young people when it comes to mobility poverty. In particular, high priority should be given to the understanding of the mobility patterns, travel behaviour and transportation challenges of young people in order to ensure the suitable availability of transport for them, which is fundamental for proper access to education, friends, social activities and, at a later stage, job opportunities.

Even if Europeans are ageing more and more and the share of young people has been steadily declining over the last ten years, the percentage of people who are less than 24 years old still represents a considerable proportion of the population as it ranges from around 30–33% in the "younger nations" (i.e. Ireland, France) to 23–24% in the "older nations" (i.e. Italy, Germany) (Eurostat 2018) (Figure 12.1).

That being said, the rise in the median age of Europeans is a direct consequence of two principal factors: a reduction in the share of children and young people in the total population (resulting from lower fertility rates and women giving birth to fewer children at a later age in life) and a gradual increase in life expectancy that has led to increased longevity. Consequently, in the last decades the share of households with children has generally

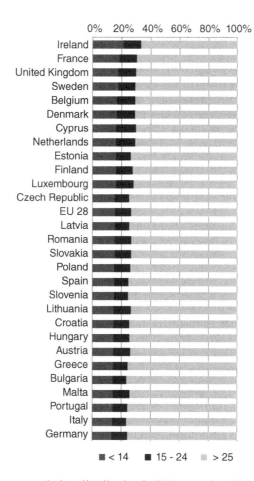

Figure 12.1 Young population distribution in EU countries, 2017.
Source: Eurostat 2018.

declined and single-person households and couples without children make up the majority of households, even with some variations among the different EU member states.

In order to describe young people's mobility, it is possible to consider different sociological and behavioural classifications for this vulnerable group. This chapter will consider the classification based on whether they are free to move independently or not:

• Children under 14 years old, who strongly depend on parents and adults to meet their mobility needs; and
• Young adults under 25 years old, who have relative freedom of movement.

Children's characteristics in transport and mobility

It is indisputable that children under 14 years old are characterised by a strong (if not total) mobility dependence on the adult world, especially that of their parents, meaning that their transport-related characteristics are highly influenced by those of the adults escorting them.

That being said, the presence of children within households has an impact on the mobility needs of the family itself in terms of the number of daily trips (typically, the need to bring children to school, medical visits, leisure and sports activities), social relationships, means of transport used and who, of the family members, is intended to carry out the accompanying role. Therefore, all these aspects have more to do with the mobility patterns related to the family life cycle than with the mobility needs of each individual.

Some interesting insights about children's approaches towards transportation can be found in theories of socialisation. It is suggested that "children learn about travel modes in the same way as other aspects of culture through agents of socialization: the family, school, media, and peer groups" and that "attitudes toward transport modes are embedded in childhood", making car dependency itself to be considered as a social problem and "tackled from a social policy rather than just a travel demand management approach" (Baslington 2008, 93–111). Also, it has been demonstrated that for children "being with and having fun with friends was a reason given for liking various modes of transport even the unpopular school buses" (Baslington 2008, 103). Children in several high car ownership households wanted their parents to change their type of car to faster models while, on the contrary, in car-free households, a higher percentage of children can imagine living happily without a car in adulthood than in households that own a car (Baslington 2008).

The above confirms the correlation between parents' car ownership and children's attitudes towards different transport modes. On the other hand, other studies confirm that the perception of various transport modes as well as the desire to drive or buy a car in the future is also influenced by peers (Haustein et al. 2009).

Other research confirms the correlation between parents' car ownership and children's attitudes towards different transport modes (Cahill 1996). According to the results of these studies, seven-year-old children already associate different modes of transportation with different levels of prestige (e.g. old people are more likely to be associated with bus travel, whereas successful-looking people are linked with car brands such as Porsche or BMW). The latter are also types of cars that children want to own when older, which can be considered as evidence of media influence on children's attitudes to transport modes (Baslington 2008).

When children grow up, their mobility pattern starts to resemble that of young adults, with some peculiar constraints that depend on the presence of a short/safe path to access school, sports and recreational activities on foot or by bike and by the presence of school bus services. Both in urban, rural and peri-urban areas, children's independence is directly related to the quality of infrastructures and the presence of services.

Challenges for children in transportation

As the picture of children's travel behaviour is complex, it is clear that the car plays a large part in the travel behaviour of children for a number of reasons including the complexity of modern life, parental perceptions of traffic danger and the possible risk of abduction, government policy giving parents the choice of school and the decentralisation of urban areas which has partly been caused by greater availability of the car. Figure 12.2 summarises the effects of modern life on children's walking and cycling.

As anticipated in the previous section, in Western Europe, children's independent mobility has seemingly declined during the past 30 years, with trends of increased car usage and accompaniment by adults (Pyer and Tucker 2014; Barker 2006, 2009). Also, it is argued that children spend an increasing amount of time on indoor activities such as computer gaming (Sandercock et al. 2012). One of the most widely cited explanations for increased dependency on parental chauffeuring is parents' concerns for their children's safety (Barker 2003, 2006; Hillman and Adams 1992). However, research in the United Kingdom has shown it is the perception of risk for children in particular that has increased and not necessarily the exposure to risk. Hence, comparing oral histories of children's everyday mobility in the past with recent accounts of children's mobility found that – apart from the perception of risk – actual mobility patterns of children remained the same (Pooley 2011).

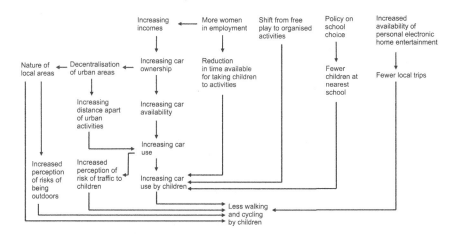

Figure 12.2 The influence of modern life on children's walking and cycling levels. Source: Mackett 2013.

If children need to be accompanied, additional traffic will be generated as parents make additional trips to drop off and then return to pick up their children. This extra traffic is likely to further reduce the likelihood that parents will grant their children independent mobility.

This strong dependence of children on adults for transportation forces households and parents to rely heavily on the use of the private car every day. This has direct effects on society in the form of negative externalities (e.g. pollution, noise, traffic safety issues) due to the increase in road traffic, particularly around highly sensitive locations such as schools or recreational facilities. In addition, negative effects on children's health such as an increase in obesity and disease due to a reduced independent mobility and physical activities are more likely to occur (European Commission 2013). Other adverse effects on children's well-being, health and personal development include loss of autonomy and access to a safe environment outside the home, lowering their quality of life and insufficient gain of practical and social skills due to inexperience in acting independently.

Further challenges may be experienced by children living in rural areas, where accessing suitable public transport might be very difficult. Lack of suitable public transport will pose restrictions on opportunities to meet with friends, push boundaries and develop independence (Hillman and Adams 1992; Romero 2010; Ross 2007), but won't help avoid built-in car dependency on growing children either (Pyer and Tucker 2014; Matthews et al. 2000; Storey and Brannen 2000; Tucker 2002).

Finally, children are a population group at high and increasing risk of poverty, due to the increasing number of low-income households and marginalised or migrant families. Such a trend is a direct result of the increase in the precariousness of the labour market and the reduction of the welfare system. This applies not only to rural areas but also to urban areas. Among the factors that accentuate the poverty of families with small children we can recognise the facts of living in a single-parent family (single adults with children accounted for 4.3% of the total number of households and this value is increasing year by year (Eurostat 2015), living in a household with a female head of the family and living in a household with a head of the family who has a low level of education.

Young adults' characteristics in transport and mobility

While children under the age of 14 who tend to be almost totally dependent on their parents, young adults start to acquire some degree of autonomy in the way they travel. However, today's young adults have also grown up with limited freedom as regards independent travel compared to previous generations, as their surrounding environment has been dominated by traffic and their travel options limited because of traffic's lack of safety (Shaw et al. 2015).

That being said, young adults are still the second group of heavy users of public transit with 36% of their trips made by this mode, after elderly people who show the highest percentage; 53% of their trips are made by public transit. They make most of their journeys by bus and use them more than the average person. They often depend on buses for access to education, training and jobs (Campaign for Better Transport 2014). In addition, they are the main group for cycling (8% of their trips by bike) and walking (12% of their trips on foot) (European Commission 2015).

Car usage is showing signs of a decrease among young adults. Compared to the mid-1990s, they make significantly fewer car journeys per person per year (377 vs. 600) and fewer young people (aged 17–20) now own cars or have driving licences (38% vs. 48%) (Campaign for Better Transport 2014). The reasons for these trends are still not well understood, but some of the insights will be discussed below.

Regardless of these numbers, the transport-related characteristics and mobility patterns of young adults are particularly difficult to frame due to the significant changes in the social and economic structure that have occurred in recent years and have strongly and rapidly modified the way in which young people live, work and travel.

Table 12.1 summarises the causal factors proposed by Delbosc and Currie (2013), Aretun and Nordbakke (2014) and IFMO (Institute for Mobility Research 2013) for the changes in young adults' travel behaviour.

In particular, two of the main developments that are having an impact on young people's mobility patterns are ICT usage as well as car usage and the perceptions of cars.

ICT and, more specifically, the impact of smartphones on young adults' mobility is relevant as such devices enable them to perform other online activities while travelling, with an impact on modal choice (Konrad and Wittowsky 2017) and because they may replace the car as the new highly visible status symbol (Tully 2011).

Specifically, in urban areas, the increasing use of ICT and the increasing availability of (travel) information is considered to make public transport and the use of new forms of mobility (e.g. multimodality and sharing) more attractive and feasible (better accessibility, the possibility of using travel time for socialising and communication).

The young "digital natives" who grow up using smartphones as part of their daily life are the main users and beneficiaries of this technology allowing them to easily access real-time travel information, download and store a ticket or unlock a shared vehicle belonging to a self-service system (e.g. a bicycle or car) (Prensky 2001). ICT obviously facilitates inter- and multimodal travel. Users can choose the mode of travel best fitting their situation, individual time and financial budget (Tully and Alfaraz 2017).

On the other hand, it is also demonstrated that the availability of information about activity opportunities as a result of ICTs and the extended

Table 12.1 Causal factors for the changes in young adults' travel behaviour

Main category	Specific factor
Demographic situation	Postponing of parenthood
	Increase in cohabitation
	Migration to a foreign country (study or work)
Living situation	Living with parents longer
	Decline in private home ownership
	Increased urbanisation
Socio-economic situation	Increased participation in higher education
	Increase in woman's labour force participation
	Increased work in the service sector
	Increase in low-waged, uncontracted work
	Decline in disposable income
Information & Communication Technologies	Increased ICT use
	Use of mobile devices to arrange everyday life
	ICT use whilst traveling on public transport
	Increase in gaming
Values and attitude	Extended youth
	Rise of pro-environment attitudes
	Decline in cars as status symbol
Transport and mobility	Improvements in public transport
	Strict driver licensing regime
	Increased car insurance costs
	Increased spending on transport
	Rise of shared mobility

Source: Based on Delbosc and Currie 2013; Aretun and Nordbakke 2014; Institute for Mobility Research 2013

scope of decision-making has an enhancement effect on two levels: it results in more trips and especially in additional longer trips (Konrad and Wittowsky 2017, 3–7). Another factor to take into account when describing young people's mobility patterns is the car's decline among this generation. Despite the fact that the acquisition of a driver's licence is still considered one of the most important travel-mode related life events (Klöckner 2004) and a sort of initiation rite, which also implies the social dimension of crossing a very important threshold into adult life (Schönhammer 1999), in many western countries car use is increasing less than before or even declining (Delbosc and Currie 2013; Millard-Ball and Schipper 2011). This is particularly appropriate for young people who seem to be less car-oriented than previous generations (Kuhnimhof et al. 2013).

The affordability of a car is a growing issue for young people (particularly the cost of insurance), parking availability is limited and maintenance costs are increasing with modern vehicle technology. In Great Britain, Norway and the United States, the decrease in licensing among young adults contributed to the recent decline in car availability (for which holding a licence is one important prerequisite). In Germany and France, where young adults' licence-holding was stable, it is mostly the decline in the car ownership of households in which young adults live that has caused the decrease in car availability (Institute for Mobility Research 2013).

There is good evidence from cross-sectional analyses of UK data to suggest that the fall in licence holding has been driven, at least in part, by changes to the lives of young people in terms of demographics, their living situation and socio-economic situation (Berrington and Mikolai 2014; Le Vine and Polak 2014).

Besides the decrease in car ownership, additional relevant developments have contributed to the observed car mileage reduction among young adults. The most important reason for young drivers' decreasing car use is that they use alternative modes more, i.e. they exhibit increasingly multimodal behaviour (Institute for Mobility Research 2013).

Moreover, according to academics such as Chatterjee et al. (2018, 29), while "economic circumstances explained the lower level of car access for Millennials living independently of other adults" for "Millennials living dependently (e.g. with parents or other older adults) there was an unexplained cohort effect" indicating a lower tendency to own or have access to a car among Millennials. Attitudinal research also demonstrates that cars are not seen as aspirational by financially better off, non-car owning young people (Thornton et al. 2011).

The car's decline in young adults' interest can also be related to the increasing prevalence of life situations which do not engender car use. The number of young adults entering higher and specialised education, located mainly in urban centres, is increasing. They usually have an urban lifestyle for their employment and social life, facilitated by modern mobile technology. In addition, young adults are increasingly adopting an environmental attitude towards transport, with public transport and walking accounting for significant shares of their modal split and cycling being regarded as more fashionable among young adults and young professionals.

Challenges for young adults in transportation

For a long time, transport-related issues came at the very top of young adults' concerns. In fact, the availability of transport is fundamental for young adults to ensure proper access to education, job opportunities, friends and other social activities. However, this is not always easy to realise, especially for those who live in rural/deprived areas or within low-income families.

Good transport links are vital for young people to access job training, but transport costs and availability can hinder both their job search and the ability of young people to remain in work if they can find it. It can also prevent access to further education and training, which in the long term has serious implications for growth and productivity.

In addition, accessing jobs by transport is not always easy for young adults, as low skilled jobs are increasingly located outside city centres where they are more difficult to reach by public transport and may involve shift or weekend work when buses are less frequent or may not run at all (Campaign for Better Transport 2014).

Another challenge for young adult's mobility is the aspect of safety in public transport. A study showed that people aged between 16 and 24 are the most likely to experience a worrying episode on public transport, especially at night (Transport for London 2014). This is particularly true for young females, an aspect that has already been investigated in Chapter 4.

Not only on public transport, but safety in general is a topic of concern for young adults. In fact, road accidents represent the first cause of death for young adults (aged between 14 and 25), with young males more exposed to this risk compared to young females. This issue is even worse for low-income young adults. In the United Kingdom, those who belong to the lowest social group appear to be five times more likely to die in road accidents than those from the highest; and "more than a quarter of child pedestrian casualties happen in the most deprived areas" (Samek Lodovici and Torchio 2015, 34).

Finally, many researchers argue that virtual mobility reduces the need for physical mobility and thus substitutes travel. E-shopping, e-learning and telework can replace the need for physical presence and hence reduce travel. In social relationships, ICT tools such as messaging and internet telephony can create a sense of proximity between people who are physically divided and thus decrease the need for physical meetings and travel (Konrad and Wittowsky 2017).

On the other hand, with growing networks and distances, the need for physical mobility to nurture these networks and fulfil social obligations is also increasing and visiting friends and relatives involving middle and long-distance travelling has become a substantial part of leisure travel and can be a significant challenge for those who are not able to conduct physical travel frequently (Elliott and Urry 2010).

Conclusions

From numerous points of view, mobility represents a fundamental aspect of children and young people's quality of life, hence the transportation needs of children and young adults must be taken very seriously by governments, local authorities and by all stakeholders involved.

So far, public policies have not always properly addressed children's travel needs, especially those related to leisure and recreational activities. The result is that, with the exception of home-school journeys (which only account for little more than a third of children's journeys), the right to have a safe travel environment outside is overlooked (Shaw et al. 2015). Such a situation is even exacerbated in rural areas, where, most of the time, adequate services to support children's travel are lacking.

To change such a trend, it is fundamental that transport authorities prioritise the consideration of the evolving needs of children and young adults more clearly and encourage bidders to improve their service offer specifically addressed to young people. In addition, a stronger impetus should be put on accessibility and the planning system should support the creation and retention of schools and jobs (particularly entry-level jobs suitable for young people) in locations that can be served by public transport. A package of appropriate measures should also be introduced to make roads safer, provide cycling training at schools and eventually reduce young people's dependence on their parents' cars (Campaign for Better Transport 2014).

In addition, it is extremely important to recognise the situations in which children also have to deal with other vulnerabilities. For example, children living in rural areas will inevitably have access to fewer opportunities than those living in central urban areas. The children of low-income families will find it harder to travel than those belonging to wealthier ones. Similarly, children of immigrants will face more severe challenges in mobility than those of the domestic population.

If youth's transport barriers and challenges are not clearly understood and properly addressed, future scenarios are likely to occur where young people without adequate access to transport services are excluded from opportunities and precluded from a bright future.

References

Aretun, Åsa, and Susanne Nordbakke. 2014. "Developments in driver's licence holding among young people. Potential explanations, implications and trends." VTI Rapport 824A. https://pdfs.semanticscholar.org/7b2f/2649063055728dee2fac2e4b3ed10dc317c2.pdf. Accessed 13 June 2018.

Barker, John. 2003. "Passengers or political actors? Children's participation in transport policy and the micro political geographies of the family." In *Space and Polity* 7, no. 2: 135–152. https://doi.org/10.1080/1356257032000133900.

Barker, John. 2006. "'Are we there yet?' Exploring aspects of automobility in children's lives." Brunel University. https://bura.brunel.ac.uk/bitstream/2438/432/1/FulltextThesis.pdf. Accessed 21 April 2020.

Barker, John. 2009. "'Driven to distraction?' Children's experiences of car travel." In *Mobilities* 4, no. 1: 59–76. https://doi.org/10.1080/17450100802657962.

Baslington, Hazel. 2008. "Travel socialization. A social theory of travel mode behavior." In *International Journal of Sustainable Transportation* 2, no. 2: 91–114. https://doi.org/10.1080/15568310601187193.

Berrington, Ann, and Julia Mikolai. 2014. "Young adults' licence-holding and driving behaviour in the UK." RAC Foundation. London. https://www.racfoundation.org/wp-content/uploads/2017/11/Young-Adults-Licence-Holding-Berrington-Mikolai-DEC-2014.pdf. Accessed 13 June 2018.

Cahill, Michael, and Winn Ruben. 1996. "Children and transport. Travel patterns, attitudes and leisure activities of children in the Brighton area." Faculty of Health, Department of Community Studies, University of Brighton, Health and Social Policy Research Centre (Report 96/4).

Campaign for Better Transport. 2014. "Why getting transport right matters to young people." www.bettertransport.org.uk. Accessed 21 April 2020.

Chatterjee, Kiron, Phil Goodwin, Tim Schwanen, Ben Clark, Juliet Jain, Steve Melia, et al. 2018. "Young people's travel – what's changed and why? Review and analysis." Report to Department for Transport. UWE Bristol, UK. www.gov.uk/government/publications/young-peoples-travel-whats-changed-and-why. Accessed 13 June 2018.

Delbosc, Alexa, and Graham Currie. 2013. "Causes of youth licensing decline. A synthesis of evidence." In *Transport Reviews* 33, no. 3: 271–290. https://doi.org/10.1080/01441647.2013.801929.

Elliott, Anthony, and John Urry. 2010. *Mobile Lives*. New York: Routledge.

European Commission. 2013. "Attitudes of Europeans towards urban mobility." Special Eurobarometer 406/ Wave EB79.4. http://ec.europa.eu/commfrontoffice/publicopinion/archives/ebs/ebs_406_en.pdf. Accessed 13 June 2018.

European Commission. 2015. "EU survey on issues related to transport and mobility." Joint Research Centre. http://publications.jrc.ec.europa.eu/repository/bitstream/JRC96151/jrc96151_final%20version%202nd%20correction.pdf. Accessed 13 June 2018.

Eurostat. 2015. *Being Young in Europe Today*. 2015 edition. Luxembourg: Publications Office of the European Union. https://ec.europa.eu/eurostat/documents/3217494/6776245/KS-05-14-031-EN-N.pdf/18bee6f0-c181-457d-ba82-d77b314456b9. Accessed 13 June 2018.

Eurostat. 2018. Database extraction. http://ec.europa.eu/eurostat/data/database. Accessed 13 June 2018.

Haustein, Sonja, Christian Klöckner, and Anke Blöbaum. 2009. "Car use of young adults. The role of travel socialization." In *Transportation Research Part F: Traffic Psychology and Behaviour* 12, no. 2: 168–178. https://doi.org/10.1016/j.trf.2008.10.003.

Hillman, Mayer, and John Adams. 1992. "Children's freedom and safety." In *Children's Environments* 9, no. 2: 10–22.

Institute for Mobility Research, 2013. "Mobility Y. The emerging travel patterns of generation Y." Institute for Mobility Research. Munich. https://www.ifmo.de/files/publications_content/2013/ifmo_2013_Mobility_Y_en.pdf. Accessed 13 June 2018.

Klöckner, Christian A. 2004. "How single events change travel mode choice – a life span perspective." Ruhr-University Bochum. https://ntnuopen.ntnu.no/ntnu-xmlui/bitstream/handle/11250/2392913/Kl%C3%B6ckner_2004b.pdf?sequence=3. Accessed 23 September 2020.

Konrad, Kathrin, and Dirk Wittowsky. 2017. "Virtual mobility and travel behavior of young people – Connections of two dimensions of mobility." In *Research in Transportation Economics* 68: 11–17. https://doi.org/10.1016/j.retrec.2017.11.002.

Kuhnimhof, Tobias, Dirk Zumkeller, and Bastian Chlond. 2013. "Who made peak car, and how? A breakdown of trends over four decades in four countries." In *Transport Reviews* 33, no. 3: 325–342. https://doi.org/10.1080/01441647.2013.801928.

Le Vine, Scott, and John Polak. 2014. "Factors associated with young adults delaying and forgoing driving licenses. Results from Britain." In *Traffic Injury Prevention* 15, no. 8: 794–800. https://doi.org/10.1080/15389588.2014.880838.

Mackett, Roger L. 2013. "Children's travel behaviour and its health implications." In *Transport Policy* 26: 66–72. https://doi.org/10.1016/j.tranpol.2012.01.002.

Matthews, Hugh, Mark Taylor, Kenneth Sherwood, Faith Tucker, and Melanie Limb. 2000. "Growing-up in the countryside. Children and the rural idyll." In *Journal of Rural Studies* 16, no. 2: 141–153. https://doi.org/10.1016/S0743-0167(99)00059-5.

Millard-Ball, Adam, and Lee Schipper. 2011. "Are we reaching peak travel? Trends in passenger transport in eight industrialized countries." In *Transport Reviews* 31, no. 3: 357–378. https://doi.org/10.1080/01441647.2010.518291.

Pooley, Colin G. 2011. "Young people, mobility and the environment: An integrative approach." In *Mobilities: New Perspectives on Transport and Society*, edited by Margaret Grieco and John Urry, 271–288. Aldershot: Aldgate.

Prensky, Marc. 2001. "Digital natives, digital immigrants." Part 1. In *Horizon* 9, no. 5. NCB University Press. https://doi.org/10.1108/10748120110424816.

Pyer, Michelle, and Faith Tucker. 2014. "'With us, we, like, physically can't'. Transport, mobility and the leisure experiences of teenage wheelchair users." In *Mobilities* 12, no. 1: 36–52. https://doi.org/10.1080/17450101.2014.970390.

Romero, Vivian. 2010. "Children's views of independent mobility during their school travels." In *Children Youth and Environments* 20, no. 2: 46–66. https://doi.org/10.3390/ijerph15112441.

Ross, Nicola J. 2007. "'My Journey to School …'. Foregrounding the meaning of school journeys and children's engagements and interactions in their everyday localities." In *Children's Geographies* 5, no. 4: 373–391. https://doi.org/10.1080/14733280701631833.

Samek Lodovici, Manuela, and Nicoletta Torchio. 2015. "Social inclusion in EU public transport." Policy Department B: Structural and Cohesion Policies – European Parliament. Brussels. http://www.europarl.europa.eu/RegData/etudes/STUD/2015/540351/IPOL_STU(2015)540351_EN.pdf. Accessed 13 June 2018.

Sandercock, Gavin R. H., Ayodele Ogunleye, and Christine Voss. 2012. "Screen time and physical activity in youth: Thief of time or lifestyle choice?" In *Journal of Physical Activity and Health* 9, no. 7: 977–984. https://doi.org/10.1123/jpah.9.7.977.

Schönhammer, Rainer. 1999. "Auto, Geschlecht und Sex." In *Erziehung zur Mobilität. Jugendliche in der automobilen Gesellschaft*, edited by Claus Tully, 141–156. Frankfurt and New York: Campus Verlag.

Shaw, Kate S., and Iris W. Hagemans. 2015. "Gentrification without displacement' and the consequent loss of place. The effects of class transition on low-income residents of secure housing in gentrifying areas." In *International Journal of Urban and Regional Research* 39, no. 2: 323–341. https://doi.org/10.1111/1468-2427.12164.

Storey, Pamela, and Julia Brannen. 2000. *Young People and Transport in Rural Areas*. Leicester: Joseph Rowntree Foundation/Youth Work Press. https://www.jrf.org.uk/report/young-people-and-transport-rural-areas. Accessed 13 June 2018.

Thornton, Alex, Lucy Evans, Karen Bunt, Aline Simon, Suzanne King, and Tara Webster. 2011. "Climate Change and Transport Choices Segmentation Model – A framework for reducing CO2 emissions from personal travel." Report to Department for Transport by TNS-BMRB and PSP. https://www.gov.uk/government/publications/climate-change-and-transport-choices-segmentation-study-final-report. Accessed 13 June 2018.

Transport for London. 2014. "Understanding the travel needs of London's diverse communities. A summary of existing research." http://content.tfl.gov.uk/

understanding-the-travel-needs-of-london-diverse-communities.pdf. Accessed 13 June 2018.

Tucker, Faith J. 2002. "Young girls in the countryside. Growing up in South Northamptonshire." PhD thesis. Sociology. University of Leicester. http://nectar.north ampton.ac.uk/id/eprint/2840. Accessed 21 April 2020.

Tully, Claus J. 2011. "Mobilisierung des Mobilen—Trends in der Jugendmobilität. Anmerkungen zur Veränderung im Mobilitätsverhalten." In *Der Nahverkehr. Öffentlicher Personenverkehr in Stadt und Region* 29, no. 7–8: 12–15.

Tully, Claus, and Claudio Alfaraz. 2017. "Youth and mobility. The lifestyle of the new generation as an indicator of a multi-local everyday life." In *Applied Mobilities* 2, no. 2: 182–198. https://doi.org/10.1080/23800127.2017.1322778.

Van Wee, Bert. 2015. "Peak car. The first signs of a shift towards ICT-based activities replacing travel? A discussion paper." In *Transport Policy* 42: 1–3. https://doi. org/ 10.1016/j.tranpol.2015.04.002.

Part IV
The fieldwork

13 Forced car ownership and forced bus usage

Contrasting realities of unemployed and elderly people in rural regions – the case of Guarda, Portugal

Vasco Reis and André Freitas

Abstract

This chapter explores the results of fieldwork activities targeting elderly and unemployed people living in Guarda, Portugal. Most of these citizens use a car, even if the former recognise that they are not fit enough to drive and the latter bear high costs. However, both groups perceive these efforts as fulfilling a basic need of living in such a rural and mountainous region where local authorities prefer to focus on urban citizens. Toll costs are regarded as a major barrier for unemployed people to get to work in territories where job opportunities are scarce, whereas elderly citizens, particularly those living in isolated places, are "forced" to use the bus and struggle with operational PT inadequacies.

Contextualisation

The fieldwork conducted in 2018 and 2019 focused on a region in the northeast of Portugal, more precisely on Guarda, the largest city in the district with the same name (Figure 13.1).

Guarda is a municipality of 42,500 inhabitants (the city of Guarda alone has about 26,500 inhabitants) scattered across a territory of 712 square kilometres and encompassing 43 smaller districts (Francisco Manuel dos Santos n.d.). Agricultural activities as well as forests cover a mountainous area, with Guarda being the highest city in terms of elevation in continental Portugal (1,056 metres above sea level). The area can be classified as monocentric, with an urban area (that is the city of Guarda) concentrating most employment opportunities, business and industrial centres.

The other main socio-economic characteristics identified by local stakeholders interviewed during the fieldwork are:

- Remote rural settlements, inhabited by middle-aged and elderly people, whilst the younger population dominates the main city of Guarda;

Figure 13.1 Map with the location of the municipality of Guarda.
Source: Wikipedia 2006 (left), TIS (right).

- Some remote-rural settlements have a higher percentage of women mostly living on their own; and
- People living in remote-rural areas tend to have low income and, hence, difficulties to afford a private car. They are often unemployed and make their living from local agricultural activities.

Target groups in this region comprise low income and unemployed people, as well as elderly people. Some of the main indicators concerning these two segments are listed below. In a nutshell, the municipality features higher motorisation rates than the average for Portugal and a much lower share of public transport usage when compared to other national medium-sized cities. The number of older people living in the region is also higher than the national average.

Table 13.1 highlights some key indicators for Guarda compared to the national average. The figures can be understood as a token of the current and future risk to social vulnerability that affects this area.

Guarda features a railway station, which is about three kilometres away from the city centre. It includes a small freight terminal and offers international services for passengers to Salamanca and Madrid as well as some domestic services to surrounding towns and to the capital Lisbon.

The region is served by some major motorways, all of which are toll roads. The availability of road infrastructure is acceptable overall, with some settlements being served by small local roads. During rough winter times, some roads are closed due to icy conditions. Public transport frequency from small villages to the city centre is limited to one or two journeys per

Table 13.1 Key indicators for Guarda compared to the average for Portugal

Parameter	Indicator	Guarda	Portugal
Transport	Motorisation rate in 2017	678.7[a]	491.2
	Modal share of public transport in 2011[b]	10.8	15.9
Elderly people	Old-age dependency ratio in 2018	34.8	33.6
	Percentage of people above 65 years old in 2018	22.8	21.7
Low income and unemployed	Purchasing power per capita in 2017	96.2	100
	Unemployment rate (%) in 2018	5.4	5.4
People living in rural and deprived areas	Population density (per km^2) in 2018	55.2	111.5

Source: Official population projections from the National Statistical Institute of Portugal (Instituto Nacional de Estatistica Portugal n.d.).

a According to the national insurance statistics which provide this indicator per municipality (Portuguese Insurance Regulation and Supervision Agency n.d.).
b According to a study about mobility in Portuguese medium-sized cities, carried out by the national transport authority (Seabra 2011).

day and becomes even more limited at weekends and during school holidays. The urban service features five public transport routes. Its taxi service declined over the past few years due to a lack of demand, both in remote settlements and even in the city, at night.

Interviews with stakeholders

The main policy makers and key stakeholders were selected taking into consideration their level of knowledge of the major transport and social dynamics and, consequently, their level of understanding of the wide spectrum of vulnerabilities felt by the locals in Guarda.[1] The five interviews involved the following stakeholders:

- The municipality (interviews were organised with traffic and education experts, as well as with the council regarding transport issues in Guarda);
- Guarda's parish presidents;
- The president of social security for the district area of Guarda; and
- A representative of an NGO providing healthcare support and transport for their members.

Even considering the variety of stakeholders, the overall outcome of the interviews is that municipality representatives and experts have *no in-depth knowledge* about transport needs in the region, especially those concerning the population that is more isolated and those of people with weaker social ties, supported by local associations and NGOs that operate in rural areas.

Among policy makers, there is great attention towards children's mobility needs, but no other groups' needs. This can be considered a knowledge issue, given that the municipality is assuming new responsibilities as a transport authority. This has forced local stakeholders to take problems more seriously, with greater care, and now vulnerable groups' mobility needs are placed at the top of the political agenda.

In addition to this, *most of the existing ongoing initiatives are still urban-centric* and designed for the urban citizen. The public transport operator (a privately owned company called Transdev) is not particularly sensitive to the needs of residents in isolated areas, focusing on a traditional, product-driven service, with poor attention to people with mobility impairments. Overall, the design of the public transport system disregards vulnerable groups' mobility needs, considering them as niche issues. Indeed, there are NGOs taking care of older citizens with reduced mobility, offering a transport service, especially to those living in very remote areas.

Transport stakeholders reported that the number of people who cannot actually fulfil their mobility needs is, however, low and often limited to villages and locations where the public transport service is not available. This also includes neighbourhoods in the city of Guarda, those located on the verge of the city and even without any public transport. Interviews with stakeholders working in social care highlighted a low number of people who are actually suffering from social exclusion. Social care stakeholders suggested focussing on unemployed persons living in sparsely populated areas as well as elderly people living in neighbourhoods on the outskirts of the city. As regards the latter group, it was suggested that people living closer to the city might be less physically isolated but more disconnected and socially isolated than those located in small villages.

Overview of focus groups: occurrences of mobility poverty

As a second stage of the fieldwork, two vulnerable segments were selected for the focus group activities: low-income and unemployed, on the one hand; and elderly people, on the other. For both groups, special attention was paid to those living in areas with limited public transport options.

Unemployed people

The focus groups revealed that unemployed people make a fairly high number of daily trips, even if the literature and especially the work carried out by Karen Lucas and colleagues have shown that the general category of people on low income tend to be much less mobile (e.g. Lucas 2012). This reasoning led one to assume that unemployed people could adapt their mobility options because they do not have regularly fixed appointments, as they are not employed. However, during the focus groups, it turned out that the unemployed people of Guarda do not use public transport at all, not even occasionally. In fact, according to official figures there is a high percentage of unemployed people who use local public transport, but in the focus groups it was not possible to interact with anyone who uses this mode of transport in the region. When asked about the reasons behind such non-use, in their view the local public transport service was of little help, expensive and unfit for people's needs.

Even though they do not currently use local public transport, participants recognise that they used to take such services when they were younger. As they perceived this experience as negative and because the service is becoming poorer and more limited, they even prevent their children from using it, picking them up from school so that they can avoid wasting time waiting for the bus to arrive.

When unemployed people need to make *long-distance trips*, interviewees mostly use rail services because a car would be too expensive and they would have to pay the toll. Toll costs are regarded as a major difficulty in moving beyond the city limits, a bottleneck for people living in remote regions where job opportunities are scarce. To overcome this barrier (and bearing in mind that public transport does not provide convenient options), participants advocate that toll costs should be different according to people's residence, allowing discounts for permanent residents.

The issue of long-distance trips is particularly important among this group. When seeking new job opportunities, they always bear in mind the costs related to transport because they are aware that extra costs besides fuel will be incurred (such as tyre maintenance). During the sessions, it was reported that some job opportunities were not accepted due to the travel costs involved. Indeed, when looking for new job opportunities, there has been a consensus among the attendees about the willingness to remain within a 40-minute journey time by car from home. Beyond that distance, total transport costs are too financially burdensome and the national minimum wage (580.00 euro) will not cope with these expenses. It was also reported that people increased this range due to a lack of job opportunities in the region (and thus accept higher transport costs which push people into cutting other budget streams for their household).

Needless to say, transport budget restrictions are a key issue for unemployed people who have limited financial resources. Nevertheless, transport – and most notably the car, perceived as the most important mode of transport – is an indispensable utility which is regarded as being of utmost importance and deserving a lot of attention. Some participants emphasise that this is the reason why they *devote about 20–25% of their available income* to car-related expenses, whilst others "prefer not to think about it, because it is really a need". It was thus made clear that unemployed people living in remote, peripheral or deprived areas often have to rely on private vehicles to access essential services, posing a substantial financial burden on their households.

Mobility needs are not limited to accessing services and job seeking, nor to other essential trips. This group of people, even if they have some material deprivation, show a great need to meet with friends and family, even if these friends and relatives live far away. However, on these occasions, the younger representatives of the unemployed group usually share their car with friends because there is no other cheap option to travel across districts, especially at night. As the use of the car severely impacts their *available budget, some constraints* arise. In order to guarantee mobility, they cut expenditure on other necessities. In fact, the younger participants said that they would like to go to the municipal swimming pools during winter and to the gym as well, but they can't because they cannot afford it.

As mentioned above, interviewees have strongly pointed out that *public transport does not fit their needs, neither its schedules nor its frequency.* This pushes them to keep a car as their dominant and only transport mode. Such a situation is considered to be without alternatives: during the focus group, participants were asked to identify barriers to using public transport and what their ideal alternative transport mode would look like. However, they had serious difficulties in answering this, bearing the current public transport service in mind.

It is also worth mentioning that the group of unemployed people has a very keen interest in new *technologies* and uses them abundantly, with the exception of the older members of the group. It was mentioned that the internet can create a sense of proximity, especially felt among those who live in isolated villages. In remote areas, where there isn't even a grocery shop, they shift to online shopping (even if the bandwidth speed is slower than the one being offered in more densely populated areas). This has decreased the need to visit shops to buy goods.

Unemployed people who live in small and remote villages and don't own a car have to count on their social network. They reported that almost everyone knows a relative or friend who owns a car and who is able to help solve minor transport issues. Due to poor public transport services, people, especially from remote villages, are willing to walk up to a four-kilometre distance. If the distance is greater, they routinely ask for car-pooling with their neighbours. In this sense, *the lack of transport options could contribute to increasing their ties with the entire community.*

To sum up, the main contribution of mobility poverty to social exclusion stems from limiting access to new jobs. In rural regions, where job vacancies are scarce, there is a need to be willing to carry out longer trips which come with a cost, especially if involving the need to pay highway tolls. Those who live in small remote villages also complained that few services remain intact, which is a disadvantage when compared to someone who lives in a city, although the impact is offset by internet access.

Elderly people

Whilst the unemployed group was more consistent in terms of its social characteristics, the elderly group was not. Elderly people who participated in the focus group can be described as having a **mix of social characteristics**. It became evident from their testimonials that those who live closer to the city centre seemed to have more cultural skills than those who live in the remote villages and have lived there all their lives.

During the focus groups, we noticed that the **distance to the city centre** is a proxy indicator of educational attainment among the elderly group, whereby those who are closer to the city show higher levels than others who live further away. This is also a disadvantage, especially regarding ICT take-up. Another outcome of the fieldwork is the **gender differences** among the elderly group, with a remarkable impact on mobility opportunity and choice. We discovered that women drive less than men, are more reluctant to use new technologies and displayed mobility needs, especially those women living in rural areas. On the other hand, a common element is that age reduces older people's ability to drive. In order to maintain an independent way of living, reduced car use makes them more dependent on other people to chauffeur them or to rely on (poor) public transport options.

Currently, most of the mobility needs of older people are met using a private vehicle. Those who do not own a car ask their relatives and friends to pick them up upon request. Sometimes, they use **public transport, but very often in combination with a private car** and there are a few times, yet more rarely, when they take a taxi. The incidence of mobility poverty is different between areas of the city and between Guarda's districts. It may be that a village is very far away from the main city, but it can still have good public transport connections due to the inter-regional bus network. However, overall, it seems that the participants living in rural areas are poorly served by public transport.

The **main mobility complaint** for reducing the use of public transport is the inadequate service frequency, often limited to one or two journeys a day. Just adding an extra journey each day, possibly at the end of the morning, may overcome this bottleneck, enabling older people to leave home in the morning, catch a bus to their destination and be back for lunch, instead of waiting for the evening bus. The current limited use of public transport also affects ticket purchasing. The price of a journey is considered too high and a monthly ticket is not always an option, also considering the limited ridership.

Possible solutions for elderly people living in rural villages are therefore quite simple. One solution would involve, first and foremost, activating a more robust public transport service to be used during Guarda's market days. This is something that participants, especially those from rural villages, would warmly welcome because they say that there is an inherited tradition of going there to buy fresh food (and, we would add, to visit the city).

Life-changing events, such as increasing cognitive problems and physical impairments, are severely impacting mobility choices, creating a more or less swift transition from driving to being chauffeured. Even when ageing adversely impacts elderly people, public transport is never regarded as a complete solution due to its strict rigidness and taxis are too expensive for regular use. So, the cornerstone of the mobility solution for people who are not fit to drive is a network of social solidarity that can guarantee transport in case of need. In other words, severe ageing and physical and cognitive impairments can hamper mobility, triggering social exclusion. The main counteracting element is a strong social network supporting these people.

Unlike unemployed people, **taxis** are the last resort, an option for very important and unavoidable appointments. Having said this, it is remarkable that almost all isolated villages *do not* offer a taxi service nowadays and it was reported that even in the city centre there are no taxis (or for that matter any public transport service) at night, namely after 11 p.m.

Finally, coming back to the **digital divide, we noticed a difference in attitude** between people living in rural and more urban areas, with the latter being more prone to using digital tools, and this applies to both women and men. Generally speaking, men, especially those who are in their sixties, living closer to an urban area, have already experienced several online interactions, including booking trips. Other groups are not acquainted with the potential of ICT and they also tend to be suspicious about it. Although not mentioned by the focus group participants, interviews with experts have shown that the incidence of mobility poverty is not increasing among the most materially deprived people due to the active role of local associations that already give some responses in terms of mobility needs.

Conclusions from the focus groups

Elderly people and the unemployed, especially the latter, do not regard themselves as being at critical risk of mobility poverty. However, it is evident that the intensive usage of private vehicles comes with a cost and is made at the expense of other social activities they value and that they are prevented from participating in these due to private vehicle dependency. For some elderly people, most notably those with fewer material resources who live further away from the city centre and the main services, they are forced to adapt to existing public transport services which they use as a last resort.

Both groups have spare time so their daily routines are similar. Those who do not work as farmers have the opportunity to help their families, chauffeuring relatives with whom they live or who live nearby to supermarkets and to different services (such as medical appointments), but also to leisure activities as well. Retired people also use the time they have available to chauffer others in need.

The presence of dense social networks seems crucial to overcoming mobility poverty and to preventing social exclusion. A tight and socially inclusive society can lessen social exclusion even when public institutions do not offer conditions for vulnerable groups to become mobility independent. In fact, considering the high costs that they have mentioned spending on their cars (accounting for almost a quarter of available income for the unemployed group, to give a vivid example of tight dependency), we can thus imagine that once public subsidies cease, they will start to become quite vulnerable.

The main results and policy implications can be summarised as follows:

– **Guarda is described by both groups** participating in the focus groups as a car dependent community. The main social and economic facilities are concentrated in the city centre and others, such as a popular retail park and the university, are located in areas that can hardly be accessed other than by car. On the other hand, the majority of the population is dispersed across the territory. Such a mismatch is connected by **car use and overuse so those who do not have access to it experience unmet mobility needs. This mobility poverty increases with age.**

– There is a shared feeling among the elderly and unemployed that public transport lacks proper investment. Buses operating in the region are depicted as those not suitable for other European cities and relocated to Guarda at the end of their lifecycle. This can trigger vulnerable groups to **move to a larger city.**

– Long-distance trips are particularly important among the **unemployed.** When seeking new job opportunities, they always bear in mind the costs related to transport. Indeed, long-distance travel is increasingly becoming a crucial prerequisite in contemporary employment (and consequently crucial to becoming employed) and disadvantages those unable to conduct physical travel frequently.

– Segments are not homogeneous and feature some important differences. Elderly people, in particular, have completely different mindsets depending on where they live, their age and cultural background. These differences matter in terms of mobility poverty. In general, **elderly people** living in deep rural and peripheral territories have to face increasing mobility problems, as their public transport offer is limited. They are quite worried about the prospect of needing a (poor or non-existent) public transport service, becoming dependent on it once they become unfit to drive.

– All in all, the **phenomenon of forced car ownership was confirmed in Guarda.** However, so far, the vulnerability of those without a car is compensated by strong social links and old and new purchase services. This boils down to itinerant vendors (travelling from village to village on designated days) as well as the increased role that online purchasing plays in this field.

All focus group participants recognise that changing mobility habits is not straightforward and requires time. Yet, they would welcome new mobility schemes provided that there are fewer rigid and more flexible public transport services, suitable to their irregular needs and flexible payment options (on subscription for example, rather than fixed monthly passes), taking into account their current socio-economic status.

Reference to the main statistical data: National Statistical Institute of Portugal (Instituto Nacional de Estatistica Portugal n.d.)

Note

1 Please see the introduction to this volume regarding the contextualisation and methodology of this research.

References

Francisco Manuel dos Santos Fundação. n.d. "Base de Dados Dos Municípios." https://www.pordata.pt/. Accessed 3 April 2020.

Instituto Nacional de Estatistica Portugal. n.d. "Statistics Portugal." *National Statistics Portugal.* https://ine.pt/xportal/xmain?xpgid=ine_main&xpid=INE. Accessed 3 April 2020.

Lucas, Karen. 2012. "Transport and social exclusion: Where are we now?" In *Transport Policy* 20: 105–113. https://doi.org/10.1016/j.tranpol.2012.01.013.

Portuguese Insurance Regulation and Supervision Agency. n.d "Insured vehicle statistics." https://www.asf.com.pt/NR/exeres/7D383D46-9431-416E-98C7-395B0A9E7080.htm. Accessed 3 April 2020.

Seabra, Maria Isabel Carvalho. 2011. *Mobilidade Em Cidades Médias.* http://www.imt-ip.pt/sites/IMTT/Portugues/Observatorio/Relatorios/MobilidadeCidadesMedias/Documents/IMTT_Mobilidade_em_Cidades_Medias_2011.pdf. Accessed 3 April 2020.

Wikipedia. 2006. *Distrito de Guarda.* Author: Rei-artur. https://commons.wikimedia.org/wiki/File:LocalGuarda.svg. Accessed 22 June 2020.

14 Perception of mobility poverty in remote peri-urban Salento, Italy

Cosimo Chiffi, Silvia Maffii
and Patrizia Malgieri

Abstract

This chapter provides a description of fieldwork activities carried out in a community of 14 small towns located in the extreme south east of Italy. This remote, peri-urban territory is included in an Italian national development programme aimed at reversing the depopulation and economic decline of more marginalised "inner areas". The research focused primarily on the mobility needs of women and people with reduced mobility. Interviews and focus groups demonstrated a high level of awareness around current mobility poverty's conditions as well as the knowledge and reclamation of more inclusive and flexible collective transport to allow personal mobility mainly on shorter routes and thus enrich living conditions within the area.

Introduction

The Cape of Leuca in the Salento peninsula is the most south-eastern part of Italy's boot in the region of Apulia. It is one of the many *finisterrae* of the European continent classified as a "remote area" by the Italian National Strategy for Inner Areas (SNAI), a cohesion and development programme addressed at territories distant from centres offering essential services and characterised by the phenomena of depopulation and degradation.

The "Inner Area of Southern Salento – Cape of Leuca" is an association of 14 small towns in the Province of Lecce (Acquarica del Capo, Alessano, Castrignano del Capo, Corsano, Gagliano del Capo, Miggiano, Montesano Salentino, Morciano di Leuca, Patù, Presicce, Taurisano, Salve, Specchia and Tiggiano) that has recently been included in the aforementioned programme with the adoption of an action plan aimed at reversing declining socio-economic trends. Four additional municipalities (Tricase, Ruffano, Casarano and Ugento) are part of the wider "strategy area" being the nearest service centres of the towns in the Cape (Figures 14.1).

This almost totally flat land is a polycentric scattered area characterised by the presence of numerous compact historical towns and just a few rural settlements. It can be classified as remote peri-urban with the presence of

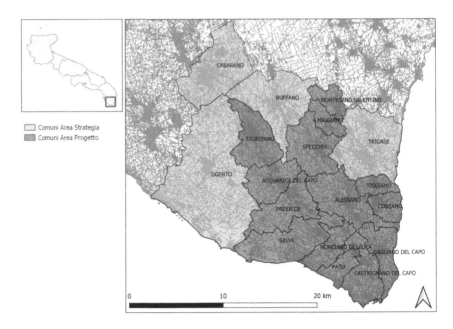

Figure 14.1 Italian Inner Areas and perimeter of the strategy area of Southern
.   Salento – Cape of Leuca.
Source: Regione Puglia – Area Interna Sud Salento 2018.

coastal marinas, inner towns and rural farms in the countryside. Built-up
areas are concentrated around town centres and the average population
density is 275 inhabitants per square km, in line with both provincial and
regional values. In terms of spatial characteristics, Southern Salento therefore constitutes more of a common rather than a peculiar peri-urban area in
Apulia (Istituto Nazionale di Statistica 2018a).

A total of 67,775 inhabitants live in the inner area distributed in little
towns and hamlets each with a total population between 1,000 and 5,000
residents. In terms of residents, the number almost doubles in the four nearest service centres. The number of residents in the inner area decreased by
0.6% between 2001 and 2011, a trend that accelerated between 2011 and 2017
with the loss of 2,176 inhabitants in parallel with a sharp increase in the
ageing index: the ratio between those over 65 and those aged 0–14 is one of
the highest in Italy ranging from 300 to 190 for the majority of the inner area
towns (Istituto Nazionale di Statistica 2018a).

The Italian National Statistics Institute (ISTAT) includes the area of
Southern Salento in the territories with a high risk and medium-high risk
of social and material vulnerability rates (Istituto Nazionale di Statistica
2018b). It shows a very high rate of unemployment especially among women

(16% to 31% across the 14 towns) and young people (23% to 52%) and a high share of elderly people living alone (22% to 30%).

The main manufacturing sector used to be what is known as the *TAC* district of textile, clothing and shoe industries. However, due to competition from other countries with lower labour costs, it almost disappeared and many companies and jobs were lost in the last 15 years. The monoculture of olive trees has recently been infected by the bacterium *Xylella fastidiosa*, a pathogen that is devastating to the landscape of Salento with severe impacts on the production of olive oil and the agricultural sector as a whole. The area is also becoming a fast-growing tourism destination with seasonal employment opportunities increasing during the summer period even if this has not counterbalanced the loss of workplaces due to the decline in manufacturing (Capestro et al. 2014).

In terms of spatial hierarchy, the city of Lecce is the closest centre to access main public offices, its university, larger employers, main events and cultural places. Inner area towns and hamlets in Southern Salento mostly belong to the main service centres of Tricase (17,000 inhabitants) and Casarano (20,000 inhabitants), where local hospitals, high schools and cinemas are located. Within the area, the towns of Alessano and Castrignano del Capo, close to the Cape, host some high schools and medical clinics. Average distances to reach basic services range between 10 and 15 km from each town whereas workplaces are more sparsely located throughout the territory and not concentrated in a few poles.

Transport characteristics in Southern Salento

The Salento peninsula is far away from national motorways which only have connections in Bari or Taranto, over 200 km and 180 km distant from the Cape of Leuca, respectively. However, the area is connected to these regional nodes by two main state road corridors also crossing the international airport and seaport of Brindisi (120 km away) and the provincial capital of Lecce (80 km away, or 70 minutes by car). Lecce is also the terminus of the national railway network. Rail corridors linking Salento to Rome and to northern Italian cities along the Adriatic line still suffer from a lack of services and some infrastructure bottlenecks, with no high-speed rail connections. The absence of faster long-distance trains and the peripheral location of the inner area therefore results in higher interconnection costs with vast and distant markets that have only been partly reduced due to recent investment and the growth of the airport in Brindisi.

The road network of the whole of Salento is more characterised by small-scale diffused infrastructures that include provincial, municipal and rural roads plus the capillary railway network of Ferrovie del Sud Est (FSE). Built in the second half of the 19th century, this state-owned non-electrified local railway has experienced a gradual lack of upgrading and maintenance of its tracks, station buildings and rolling stock in the last decades. The absence of proper safety systems resulted in a maximum permitted speed of only 50 km/h (Agenzia Nazionale per la Sicurezza Ferroviaria 2016)[1] making travel times very long on almost every connection and a load factor below 15%. In 2016, due to financial and judiciary problems experienced by the company, the network was taken over by the Italian railways (Gruppo Ferrovie dello Stato Italiane – FSI) and a big renovation programme has finally started. The FSE network is therefore in a transition phase with several projects, including its electrification, currently ongoing (FS News 2019).

The local bus network is not optimised nor adapted to the dispersed territorial context with many overlapping bus lines (served by a fragmented range of different companies), no demand-responsive services being operated in combination with PT corridors and existing services targeted principally at home-to-school trips. Most towns in the province of Lecce are served by just two bus journeys per day and only during the summer period does a seasonal bus service called Salentoinbus with higher frequency provide public transport accessibility to the towns and coast in order to support the tourism industry.

Due to the low mobility demand, but also the poor quality of public transport services, many commuting trips in Southern Salento are made by private motorised transport: 61–66 per 100 inhabitants according to 2011 Census data in the different towns. This has also been confirmed by a recent mobility survey carried out at regional level (ASSET Regione Puglia 2019).

The motorisation rate of 57.7 cars per 100 inhabitants in the inner area is a bit lower than the national average (62 cars/100 inhabitants), but higher than Apulia's regional rate (54 cars/100 inhabitants) (Automobile Club d'Italia 2018).

Interviews with local actors and stakeholders

The social groups primarily targeted by the fieldwork, conducted in 2018 and 2019, were women and people with reduced mobility.[2] In preparation of the focus groups with citizens, several key stakeholders have been interviewed, ranging from educational and social institutions to civil society associations, from local politicians to transport suppliers' representatives. This offered some insights into the local context and the mobility situations with which the two vulnerable categories have to contend every day, allowing an initial understanding of the mobility system as a whole (intercepting problems also affecting other social groups, particularly the elderly and young people).

Women's mobility-related situations and mobility needs have been discussed with the representatives of Le Costantine, a local foundation providing training and working opportunities for women in biodynamic agriculture and handmade weaving (but also running a Steiner school and educational centre for young people). We met Serenella Molendini, currently Deputy National Advisor on Gender Equality (and formerly advisor of the Puglia Region and the Province of Lecce) as well as the social cooperative Comunità di San Francesco that runs the local anti-violence network. The Comunità di San Francesco also offers social and educational services to elderly people, children, families and people with disabilities (which is also of interest to the second focus group).

In order to understand the mobility needs and available transport options for citizens with reduced mobility, the project interviewed officers at the Local Health Service ASL, the University of Lecce and the operators of Terra Rossa, a social cooperative working in both physical and cognitive accessibility. In addition, for both vulnerable groups, the footwear manufacturer Calzaturificio Sud Salento (a big employer in the inner area with 250 workers), the railway company Ferrovie del Sud Est and some mayors have also been involved in order to include the points of view and perceptions of mobility poverty conditions in the area by some key actors and policy makers.

Main outcomes of the interviews

The overall perception among the stakeholders not providing transport services and the politicians is that the local collective mobility system is almost "non-existent, inefficient and of very poor quality". They identify mobility as a crucial issue because it cuts across different parameters such as access to work, education and leisure opportunities for inner area residents. With reference to the two vulnerable social groups, they underlined the impacts of the mobility system on family care and independence, allowing or preventing a work-life balance both for women and for families who need to assist a person with reduced mobility.

The remoteness and peri-urban characteristics of the area are ambivalent. On the one hand, they mean longer commuting trips to reach Lecce and other larger service centres; on the other hand, they demand more frequent, often shorter trip chains (particularly for women) between two or more towns to access facilities for children, the elderly and disabled care as well as social activities. For longer distances, there is no alternative to the private car, with the rail service still being too slow (as mentioned earlier), while the bus network is mostly organised around the needs of public schools (but not other training and educational centres), only offering morning journeys.

In Puglia, we have one of the most advanced legislation and policy frameworks addressing the needs and well-being of families – argued Serenella Molendini. The whole system of social services – from kindergartens to socio-educational centres for disabled minors – has been restructured with more diversified and high standard facilities that are now also closer to people and towns.

These structures are fundamental especially in an area where work resignations following maternity leave are still too high and unfortunately not declining. The gender equality advisor pointed out the efforts made using the so-called Piani dei Tempi e degli Spazi (Territorial Time and Site Plans), a tool for analysing and planning physical accessibility and coordinating opening/closing times for certain services and activities in order to meet family (i.e. often women's) needs and work constraints. This tool was very useful for the acquisition of knowledge on mobility poverty conditions in the region, but unfortunately led to no implementation nor adaptation of services in order to fill the gaps:

> There are no virtuous examples of mobility services that have joined the new reorganisation of family care services so the transport barrier remained unsolved. Especially in small towns, the hourly flexibility of workplaces and public services is not as important as the public transport links with neighbouring towns and with larger municipalities, which are almost nonexistent.
>
> added Molendini

This situation was also confirmed by the stakeholders offering educational and training opportunities for the communities of Southern Salento. They are often forced to organise door-to-door minibus services themselves, thus reducing the budget for other courses or the number of participants. Commuting trips to reach workplaces are often self-organised with informal carpooling teams by the employees. In this respect, these bottom-up actions are an answer to very poor mobility options and they can also be a foundation for future activities.

Special needs transport for primary school pupils is organised directly by the municipalities, whereas secondary school students with reduced mobility can use dedicated services (established by the Province of Lecce). The University of Salento also has a dedicated minibus, but such a service can only offer one return trip per week due to the limited coverage and capacity of the service, particularly for longer distances. The person in charge of the Integration Office at the University of Salento, Paola Martino, underlined the fact that

> every attempt to coordinate the available public transport offer failed in the past because, despite the limited but rising number of buses equipped with ramps for wheelchairs, the operators are not able to guarantee their usage on a certain line or ride.

This is mostly because vehicles are assigned to services according to drivers' shifts and not to lines, thus failing to guarantee a proper accessibility and quality standard. The Public Health Service ASL provides special transport services throughout the whole province with 18 equipped minibuses allowing people with reduced mobility and disabilities to reach hospitals, rehabilitation services, mental health and socio-educational centres. Despite the availability of the vehicles, as well as the coverage of a large part of both fixed and running costs associated with the service, these minibuses cannot be used for any other travel purposes.

Focus groups in the inner area of Southern Salento

The fieldwork activity in Southern Salento ran two focus groups, with parallel and plenary sessions, devoted to vulnerable group categories. These were held in the small town of Patù (1,700 inhabitants) and in the main city of Tricase. In total, 17 women and 28 people with reduced mobility took part in the two events. The second group constituted, for the most part, people with permanent reduced mobility, but also elderly people with walking difficulties and a woman with a broken leg. Parents and representatives of local associations for disabled people were also present.

The involvement and participation efforts of people in wheelchairs were particularly challenging: they heavily rely on the availability of their parents and relatives to chauffeur them by car. Preliminary contacts by phone revealed the need for many of them to establish a very precise agenda for their daily and occasional mobility needs in advance, depending on the agendas of their relatives and friends. A dedicated special transport service to reach the venues was therefore also offered by the focus group's conveners, but only one participant chose this solution. Others initially asked about the presence of a professional assistant but in the end preferred to use the more familiar option of being supported by relatives and friends. Two invited disabled participants cancelled their attendance due to last minute changes in the availability of their relatives. Participants based in the village of Patù came with their own motorised wheelchairs; one disabled participant drove his own specially equipped car.

At both events, people with reduced mobility were quite heterogeneous in terms of age, working situation and level of autonomy. This heterogeneity was less marked in the two focus groups of women: most of them were employed or self-employed and married, though from different generations. In both focus groups, the level of attention and participation was particularly high. Women's focus groups actively debated topics such as access to work and education, family care, personal fulfilment and more generally the so-called "work-life balance", that is, the difficulties in coordinating work and personal life.

Focus groups were carried out following a participatory method inspired by GOPP (Goal Oriented Project Planning) according to an established approach: the emergence of problems (first discussion), definition of problems (by using sticky notes) and clustering of problems. A white message board with the title "Which mobility problems do you face in your daily life?" was used to structure the discussion. Within their focus group, the two categories of vulnerable users have been able to identify and define hetero-geneous problems related to their mobility experiences as described below.

Women's travel requirements, mobility poverty and social exclusion

When discussing women's mobility problems in Southern Salento, the first issue raised by the participants was naturally the absence of proper public transport options. This was seen as one of the main barriers affecting wom-en's capability of moving around, preventing them from being autonomous and making them more prone to social isolation. Without owning a car, many places in the whole province are in fact considered inaccessible.

Immigrant women in particular who need to reach local health services or who are users of the anti-violence centre in Ugento are highly affected, as attested by two female workers participating in the focus group. Indeed, a second car is not available in every family and the one the family owns is mostly used by the male adult. However, women who can use their own car also expressed dissatisfaction: cars can be a convenient means of transport, but the participants reported that they were "forced" to use (and overuse) private motor vehicles. In both focus groups, participants asked not only for an improvement of the public transport options (e.g. more frequent and faster trains or express buses to reach destinations further away), but im-mediately identified more flexible, on-demand minibus services as the most appropriate solution for travel within the inner area. This was somewhat surprising to the facilitators, considering this type of service is not present in the region. On different occasions, they declared their willingness to also use commercial ride-hailing options if available:

> I use the car for both commuting to work and for leisure. I would defi-nitely prefer to use it less. Now that my leg is broken, I'm constantly dependent on someone else from my family and this is a total waste of time for them. A call-a-bus service with minibuses would be ideal.

Women struggle to coordinate trips to accomplish work, family and leisure activities with men usually providing very little support. A lot of time has to be invested to satisfy all these mobility needs and to cross and reach towns that are relatively close to each other. Also, as stated by one focus group member, "if you drive a lot for work purposes and for family reasons, you tend to avoid additional car trips for leisure and social activities (i.e. to meet friends)".

Another issue that greatly affects women is their limited number of op-portunities in terms of jobs and higher/professional education, which often require a long commute such as to reach the city of Lecce. The absence of proper public transport can seriously compromise women's future job and training opportunities and personal growth. Safety and security issues are always a concern for women in transportation due to poor road and signage conditions as well as due to the absence of proper lighting that create difficulties in driving across towns or along secondary roads. In addition, the absence of police checks contributes to the creation of a sense of danger as does the lack of public transport options that force young people to use private cars, fostering anxiety, stress and concern among mothers.

People with reduced mobility needs, mobility poverty and social exclusion

For this category, the absence of accessible public transport alternatives or dedicated mobility services was quite immediately identified as the main element affecting people with reduced mobility autonomy and social life. They also indicated the absence in the area of a special taxi/equipped hire car with driver services (only one local entrepreneur offers such a service!), pointing out how a similar system addressed tourist's needs in the summer season. It was also reported that most buses and trains are not equipped with lift platforms making them therefore inaccessible to people in a wheelchair. Public transit personnel are neither adequately trained to assist them (e.g. some do not know how to lift the platforms) nor to provide travel information. Any information acquired is simply confusing and unclear.

This lack of coordination and reliable information hamper mobility, especially when the trips are made in two or more parts: "if from my town I have to get to Lecce and I have to change bus or operator on the way, then I am not sure I can access the second vehicle because accessibility standards are not guaranteed". Public transport services are thus perceived as inadequate. Quite often public transport facilities are planned without considering the requirements of persons with a disability and, even when the equipped vehicles exist, quite often stations and bus stops are not accessible. Services and equipment cannot be used independently and without assistance, which is also a barrier.

The overall perception is that public institutions are not sensitive towards the basic needs of people with reduced mobility: "The disabled person seems to be a burden for the municipal administration" There is no monitoring and evaluation of the accessibility requirements of city services, buildings and especially transport infrastructures: many municipalities have not adopted the PEBA (Plan for the elimination of accessibility barriers), nor nominated a representative to defend the rights of disabled people. Many focus group participants declared they would be available to take up this role.

Universal design principles and accessible public transport solutions were well known among the participants as well as their potential benefit for the whole community: "In some cases, there is a lack of adequate knowledge from the municipal administration. Some services may also be useful for other categories of users". "In many European cities, there are integrated services that are good for all citizens even with beautiful aesthetic solutions and the staff in these cities receive comprehensive training."

Additionally, leisure facilities (like beaches and festival venues) are not accessible to such vulnerable users: available services are not provided for these events, which de facto excludes them from public activities and society at large. To compensate for these disadvantages, both local associations very often provide mobility services on a volunteer basis and for free.

When dealing with the use of private cars, both as a passenger or driver, the absence of dedicated parking spaces is perceived by users as a key barrier even to travel within the same town: "Finding a car park is a utopia in my town. I stopped using the car, I prefer to go around in my wheelchair, but I encounter many architectural barriers". Car parks are not only located too far away from the final destination, they are also inaccessible, as no proper attention is paid to basic design elements. These facilities are designed with insufficient space to allow a disabled passenger to get out of a car, with assistance from another person or by themselves, and frequently these spaces are used by non-authorised drivers, with no enforcement to prevent illegal parking.

An important aspect that emerged from the discussion is that people with reduced mobility focus on door-to-door accessibility. For them, infrastructures and means of transport are all part of the same "problematic" system: whatever the barrier, the impact is reduced mobility, irrespective of whether it is an interrupted pavement, lack of parking space, absence of equipped vehicles or absence of ramps. Such a systemic and inclusive approach offers us important insights.

Conclusions

Sustainable mobility and – in particular – public transport have not been at the top of political agendas for many years. The remoteness of Southern Salento and its large dependency on private transport only recently prompted local policy makers to focus on the topic. The main outcome has been the "Inner Area Strategy" report, which focuses primarily on the topic of sustainable mobility. We can find some signs of a paradigm shift: gradually, the political debate is moving away from the subsidising of improvements and investments in road infrastructure (including additional high-capacity state roads) to a higher level of service in collective public transport. The role of local railway and a more coordinated and efficient bus and rail network is getting more attention such as innovative solutions like flexible on-demand minibus services. These new approaches appear to be in line with the perceived needs of local communities. Even if no tangible changes are visible so far, investments and public funds for the implementation of these new plans have been secured by the Inner Areas National Committee and the Apulia Regional government.

We also noticed a lack of appropriate knowledge about the term and meaning of mobility poverty, although almost all interviewees and focus group discussions recognised the issue. Overall, the current public transport system largely inhibits the possibility of accessing public services, social facilities as well as job/professional and leisure opportunities for the two social categories investigated as well as other vulnerable groups like the young and elderly people or immigrants without a car.

The fieldwork activities in Southern Salento show that in a remote rural and peri-urban territory, in addition to faster and more efficient opportunities to link with far away destinations, communities are requesting more sustainable and flexible local collective transport services, those allowing mobility on shorter routes in order to enrich living conditions within the area. In this regard, coherent and coordinated transport service design and inclusive mobility are seen by both focus groups' participants as important tools, having the potential to improve significantly the mobility of vulnerable groups. We also found an acceptance of moving away from personalised (and thus limited and expensive) services towards integrated solutions, suitable for a large pool of users and their needs. Interviewees and focus groups' participants were confident that these innovative services will also positively impact other vulnerable groups like the elderly, children and young people and immigrants.

Notes

1 This is a national restriction issued by the Italian railways safety agency ANSF (Agenzia Nazionale per la Sicurezza Ferroviaria) on all rail infrastructure where a train protection system is not operational. The measure was taken in 2016 after a train collision happened in northern Puglia that caused 23 deaths. ANSF Circolare 009956/2016.
2 Please see the introduction to this volume regarding the contextualisation and methodology of this research.

References

Agenzia Nazionale per la Sicurezza Ferroviaria. 2016. "Pubblicata la nota ANSF 009956/2016 del 26 settembre 2016 relativa ai Provvedimenti urgenti in materia di sicurezza dell'esercizio ferroviario sulle reti regionali di cui al decreto 5 Agosto 2016 del Ministro delle Infrastrutture e dei trasporti 'Individuazione delle reti ferroviarie rientranti nell'ambito di applicazione del decreto legislativo 15 Luglio 2015, N° 112 per le quali sono attribuite alle Regioni le funzioni e i compiti di programmazione e di amministrazione'". https://www.ansf.gov. it/documents/20142/241502/0099562016_260916.pdf/66f36027-5d40-2b4b-902d-38408d6f8612?version=1.0&download=true. Accessed 19 September 2020.
Automobile Club d'Italia. 2018. ACI Autoritratto. http://www.aci.it/laci/studi-e-ricerche/dati-e-statistiche/autoritratto.html. Accessed 12 December 2019.
Capestro, Mauro, Elisabetta Tarantino, Fabrizio Morgagni, Eleonora Tricarico, and Gianluigi Guido. 2014. "Distretti calzaturieri in crisi: cause del declino e

strategie di rinnovamento." In *Economia e Società Regionale* 1: 187–212. https://doi.org/10.3280/ES2014-001015.

FS News. 2019. "Affidati i lavori per l'elettrificazione della linea Martina Franca-Gagliano del Capo." 31 July 2019. http://www.fsnews.it/fsn/Gruppo-FS-Italiane/Altre-societ%C3%A0/Ferrovie-del-Sud-Est-affidati-i-lavori-per-elettrificazione-della-linea-Martina-Franca-Gagliano-del-Capo. Accessed 12 December 2019.

Istituto Nazionale di Statistica. 2018a. I.Stat. http://dati.istat.it/. Accessed 12 December 2018.

Istituto Nazionale di Statistica. 2018b. 8milaCensus. http://ottomilacensus.istat.it/. Accessed 12 December 2018.

Regione Puglia – Area Interna Sud Salento. 2018. Documento di Strategia approvato. http://old2018.agenziacoesione.gov.it/it/arint/Strategie_di_area/Strategie_di_area.html. Accessed 12 December 2019.

Regione Puglia – Assessorato al Welfare, Unità Referente Pari opportunità e non discriminazione. 2018. "I Piani dei Tempi e degli Spazi." http://www.pari opportunita.regione.puglia.it/piani-dei-temi-e-degli-spazi. Accessed 12 December 2019.

Regione Puglia – ASSET. 2019. "Indagine sulla mobilità dei cittadini residenti nel territorio regionale finalizzata allo studio della domanda di trasporto in Puglia." http://asset.regione.puglia.it/?mobilita-indagine. Accessed 12 December 2019.

Urso, Giulia, Marco Modica, and Alessandra Faggian. 2019. "Resilience and sectoral composition change of Italian inner areas in response to the great recession." In *Sustainability* 11, no. 9: 2679. https://doi.org/10.3390/su11092679.

15 Isolation, individualism and sharing

Mobility poverty in Naxos and Small Cyclades, Greece

Akrivi Vivian Kiousi, Mariza Konidi and Dariya Rublova

Abstract

This chapter presents the results of the fieldwork carried out in remote and secluded areas on Greek islands through stakeholders' interviews and focus groups with citizens. The main objective was to assess groups vulnerable to mobility poverty, in particular children and populations living in remote rural areas. This was implemented on the island of Naxos and on the much smaller one of Iraklia.

Interviews and focus groups show profound isolation, but also robust commitment to seeking collective solutions to mobility poverty, at least on Iraklia. While privately owned cars are the dominant means of transport, informal peer-to-peer car-pooling is very common in remote areas. However, there are weaker social strata, as children, the elderly and unemployed are more prone to suffer social exclusion.

Introduction

Big Sweet has this island, virtuous are the faces of people, piles are shaped by melons, peaches, figs and the sea is calm. I looked at the people – never this people have been frightened by earthquakes or by Turks, and their eyes did not burn out. Here freedom had extinguished the need for freedom, and life spread out as happy sleeping water. And if sometimes was discomposed, never rose tempest. Safety was the first gift of island that I felt as walking around Nàxos.

(Kazantzakis 1965)

The location: Naxos and small Cyclades

This chapter presents the results of fieldwork, as conducted in 2018 and 2019, in remote and secluded areas on Greek islands through stakeholders' interviews and focus groups with members of socially vulnerable groups.[1] Our main target groups were (i) people living in remote rural areas and (ii) children. The activities ranged from interviews with stakeholders to focus groups run on Naxos (a bigger island with considerable infrastructure (health, education) and Iraklia (a smaller island). The fieldwork was conducted on the island of Naxos and on the much smaller one of Iraklia including the target groups mentioned above.

Fieldwork localisation

The Cyclades differ significantly in size, geography and historical background. The name Cyclades comes from the fact that the group is (roughly) the shape of a circle.

Naxos and Small Cyclades is a municipality in the Naxos regional unit, South Aegean region, consisting of the island of Naxos (the largest of the Cycladic islands) and the surrounding smaller islands of the Small Cyclades: Donousa, Iraklia, Koufonisia, Schoinoussa and several smaller islands. The municipality of Naxos covers an area of 495.76 square kilometres and has a permanent population of 18,904 inhabitants (7,070 of whom in Naxos town, also named Hora) (European Commission 2019). For the archipelago, the end of the Second World War was the beginning of migration – all in search of a better life – to Athens and other major Greek cities as well as other countries. It was during this period that the villages of Naxos lost a large part of their population, as did other islands (2019) (Figure 15.1).

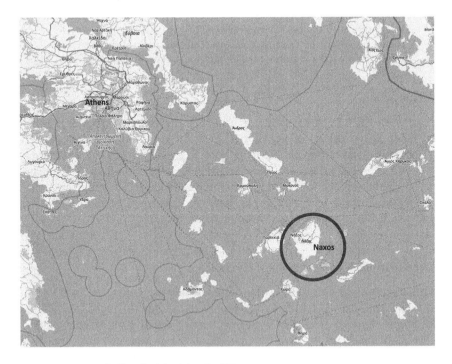

Figure 15.1 Map indicating the island of Naxos.
Source: © The contributors of OpenStreetMap, openstreetmap.org

Naxos

Currently, the island of Naxos is predominantly a rural region with a dispersed settlement (22 villages, ranging from 10 to 2,000 permanent residents). The population also extends into the more mountainous locations as well, where people are mainly involved in agricultural activities. Hora (the main town) is the main harbour and administrative centre of the island, a town which has managed to maintain a sizeable population. The coastline is an important tourist destination during the summer (Figure 15.2).

By contrast, some of the mountain villages are uninhabited or may have very few inhabitants, mostly of the older generation. Other residents commute seasonally between the village and the central island. Other inland

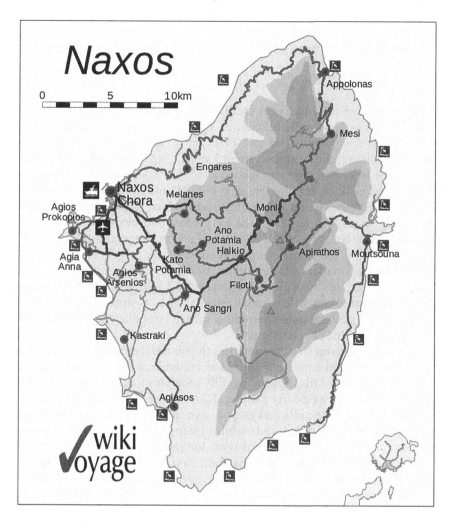

Figure 15.2 Map of Naxos.
Source: Wikivoyage 2012

villages are bigger, with a population of up to 2,000 inhabitants, with a considerable number of children. Until a few years ago, Naxos' mountain areas were unknown to tourists, but more recently many villages are making efforts to attract tourists, especially during the high season, featuring local festivals, the production of local crafts and food specialties. The fieldwork addressed people living in the mountain areas, the most disadvantaged in terms of accessibility, considering the distance of up to 45 km to the main settlement of the island, i.e. Naxos town.

Transport conditions are difficult; transport and travel are often a strain on the energy and resources of the rural poor. Very high car use and ownership have been recorded both on Iraklia and inland Naxos. The end of the tourist season also means reduced public transport trips for residents, thus making it difficult for non-drivers to be transported. The road infrastructure is poorly maintained. Streets are fairly good around Naxos town and the seaside touristic areas, while they are much worse in the mountains.

Iraklia

As mentioned above, Iraklia is a small island, with merely 141 permanent inhabitants on its 17.8 square kilometres. Iraklia is positioned about 18 miles from Naxos town, equivalent to 1 hour and 15 minutes by ferry boat. Its population is split between two small hamlets: Agios Georgios (where the island port is located) and Panagia, located four kilometres from the port. The highest peak on Iraklia, Papas, is only 418 metres high. Most beaches on Iraklia are rocky and can be reached by boat. Iraklia has no public transport coverage (apart from the ferry boats and bus operating only in the summer to cover the needs of tourists) and poor social infrastructures. While Naxos, as the bigger island, has considerable infrastructure (health, education), the smaller ones depend on the bigger island for many services.

Some of Iraklia's background situation and fieldwork results can easily be replicated on other small islands with a similar geographical background (mountainous footpaths, varied coastline) and spatial and socio-demographic characteristics.

The isolation factor that characterises the islands (being areas surrounded by the sea) is often combined with land remoteness, factors that create a "double" isolation. This triggers mobility problems for different vulnerable social groups such as elderly people, children (due to limited educational, social, health service units and large distances, e.g. a long way to school), women and people with reduced mobility (inappropriate infrastructures, inadequate/non-existent public transport, long distances). The geographical restriction of the islands is also the main characteristic that differentiates them from the continental remote areas. In this vein, we found differences compared with similar continental areas, but also between big and small islands in the same island complex, and, naturally, within different areas on the same islands.

Why Naxos and Iraklia?

The running liaisons with CIVINET CY-EL (a network of around 120 local authorities) and its engagement in transport-related activities made it the perfect partner to conduct fieldwork in the Cyclades (Civitas 2019). After thorough discussions and desk research, the choice fell on the island of Naxos and on the smaller Iraklia, the latter being a very good example of a secluded case, affected by seasonal mobility disparities. Fieldwork was supported by the local authorities providing access to local stakeholders. The two islands display difficult access, geographically secluded areas and a remarkable seasonal difference in mobility needs due to the arrival of tourists in the summer.

The interest in the fieldwork by the Naxos and Small Cyclades local authority was also taken into consideration during the liaison phase when the municipality was briefed on the project. Naxos local authorities were also interested in the study in order to obtain additional information in elaborating their first Sustainable Urban Mobility Plan.

Naxos and Iraklia

The main income for the local population, before the rise of tourism, was derived from farming and animal husbandry, both of which are well developed. In the past, many of the villages used to have a leading financial role in the island's economy. In this regard, most of the island's population lived in the inland villages and Naxos town was not the dominant centre it is today. Both study areas present a rapid development of tourism which kicked off in Naxos around the 1980s, with the influx of tourists reaching around 500,000 visitors each year. Below, we present the main identified socio-economic characteristics of the wider region for which official statistics are available:

- Level of education (Regional Unit Naxos (Eurostat 2018)): university degree 11.4%, holders of a high school diploma or those who have completed vocational training 24.4%;
- Unemployment rate (Regional Unit Naxos): 15.2%;
- People at risk of poverty or social exclusion (Aegean Islands and Crete): 37.5% in 2016 (in comparison, total for Greece: 35.5%);
- GDP per capita in PPS (EU28=100), South Aegean (including Naxos): 21,900 euro/year; and
- More than 90% of the GDP in South Aegean Islands comes from tourism-related activities.

Indeed, the tourism industry is the main source of income and development. According to the fieldwork findings, it seems that the quality of life has been improved due to this tourism development. An increase in the number of permanent residents has also been openly reported, especially in the more touristic old seaside villages. On the negative side, this tourism is unsustainable. At the peak of arrivals, there are issues regarding pollution, car traffic and parking. A similar situation can be found on the Small Cyclades islands.

However, things are very different for the inland and mountain areas, which so far attract almost no tourism and, thus, do not benefit from similar economic growth. Agricultural activities, which characterise these regions, bring inadequate incomes and there are almost no other work opportunities. As highlighted by some of the interviewees, ad hoc initiatives to support local development, run by residents and local authorities, have so far not helped to improve the situation. Not surprisingly, many residents of the mountain areas leave their villages, shifting their professional activities from agriculture to tourism, which is concentrated on the coastline. With regards to the availability of infrastructure and services in the mountain regions, the villages usually have a mini market, a pharmacy and a rural doctor. Other services, supermarkets and banks are available either in a big village or in Naxos town. There is one ambulance on the island, but it has to cross long distances (up to 45 km) along bad mountain roads to reach the patients in the mountain areas.

Local actors and key stakeholders

The fieldwork started with selective consultation of key local players and inhabitants. The combination of desk research and in situ sessions preliminarily aimed to define the current situation, what remains unclear and what needs further analysis. The link between mobility poverty and social exclusion was evident and, ultimately, the fieldwork intended to assess the current challenges experienced by the inhabitants. The engagement of local stakeholders and discussion with them went beyond transport matters, covering a larger pool of topics: for example, the needs of pupils due to increasing school enrolment rates, access to health services by the inhabitants, female empowerment and improvements in agricultural productivity. According to this approach, the following stakeholders were reached and their opinions recorded:

- The Municipality of Naxos and Small Cyclades;
- Public Bus Company (KTEL);
- Taxi (companies) owners;
- Sea taxi owners;
- Local ferry boat line connecting Naxos with Small Cyclades islands (subsidised by the state to keep the Small Cyclades islands connected even in the seasons with very low influxes of passengers);
- School authorities (primary and high school principals); and
- The Centre for Creative Work for People with Disabilities.

Consolidated results

First, we can detect a causal link between poor transport services and the tourism industry, since per occasion and region the coverage of offerings is not satisfactory at all times and seasonally dependent as expressed by different individuals mentioned earlier. In addition, we encountered members of lower socio-economic strata that require special attention regarding service offers to be assisted and have equal opportunities with the rest of the islanders who reside in more central and populated areas of the islands.

We should also report a remarkable difference in mindset between Naxos and Iraklia. As discussed in the focus group talks, people living inland, and specifically in mountainous Naxos villages, are hindered from cooperating with each other and also reluctant to participate in collective actions since they are of a different mindset, but are also worried about the risks associated with new ideas and innovative solutions. By contrast, in Iraklia, residents act collectively and more as a team. Overall, it can be derived from the study that residents are eager to find collaborative and innovative solutions and are more driven by the value of the result than putting too much weight on the challenges and risks associated. The latter is not meant to undermine their responsible side, but they tend to be more decisive. Of course, it is easy to see how in Iraklia the community of permanent residents is really small: their profoundly isolated lifestyle, the long winter months and very limited number may explain the need to face challenges together and maintain strong social ties. This applies not only to native people, but also to those who moved to Iraklia in more recent years.

The difference in attitude was also observed in relation to those with a disability (especially people with a mental impairment): While on Naxos families often hide their (mentally) disabled relatives at home, in Iraklia people with disabilities tend to be part of society to a greater extent. However, both on Naxos and Iraklia, informal car-sharing and chauffeuring by friends, neighbours and relatives is popular and fills many transport gaps.

Populations living in remote rural areas

On Naxos, the population living in the remote mountain areas usually faces difficulties in their daily mobility since service offers are affected by seasonality. As public transport cannot meet their on-demand needs and schedules are not satisfactory, as they expressed, most of the adult population (both men and women) rely on car ownership, which is extensively used for their mobility needs and as a means to an end to assist their social inclusivity. Having said that, the problem is more severe for those who do not own or cannot drive a car (i.e. the elderly, young and those on low income) since they experience poor socialisation: their prospects are restricted to the people and opportunities provided by their village. One way of overcoming this is informal car-sharing as practised especially in mountain regions. The peer-to-peer, informal service as stated during the discussions helps people with no car ownership and it is mainly offered by neighbours and relatives on a voluntary basis. However, apart from basic needs, other needs are left unmet. Job opportunities are also restricted by people's ability to be mobile. Thus, the combination of the remoteness of the location, insufficient coverage by public transport services and the high price of tickets are the main factors depriving many inhabitants of broader opportunities and socialisation.

Another category of people who can be considered socially excluded are disabled people, especially those with some sort of mental disabilities as we recorded during our visit. Being different is still stigmatised. Physically disabled people have difficulties in moving around the island due to the total absence of the necessary infrastructure, leading to their isolation from the rest of society, since their needs, which are more specific, are insufficiently met demonstrating a lack of specially designed services.

Although the whole population suffers from substantial isolation due to the topology and very limited design of the services offered by the state, permanent residents on Iraklia do not feel socially excluded. They happily support their decision to reside on the island knowing the limitations and difficulties involved with this decision. What was surprising was that the social and mutual support ties are strong and social gatherings and engagement are taken very seriously on the island.

Children

During our study, children in remote mountain areas expressed that they face the consequences of the remoteness of their villages. However, as the children noted during the focus groups, school buses are subsidised by the regional authority and are free of charge to the pupils. The bus passes through many villages collecting pupils, but these routes also include the most remote places often resulting in a rather time-consuming ride (up to one hour). School buses very frequently report delays and their schedule is aligned only with the school timings, making them not suitable enough to be a worthwhile choice to support their extracurricular activities. Children mentioned they need to cover even longer distances (up to 45 km) to reach their after-school activities, which in the majority are concentrated in Naxos town. The remoteness, the necessity to travel long distances and limited transport options result in tiredness, difficulties in joining after-school activities and, as a result, they frequently abandon their plans. This leads to a lack of socialising. Despite the difficulties they face, many of the residents of mountain regions do not want to permanently change their place of living. Only the youngest showed a willingness to abandon their village and even the island in order to follow their study and future carrier paths.

Conclusions

The main outcome of the fieldwork confirms that transport infrastructure and available services enable people to develop their social and daily life. Very high car use and ownership have been recorded both on Iraklia and inland Naxos. The road infrastructure is poorly maintained. Streets are fairly good around Naxos town and the seaside touristic areas, while they are much worse in the mountains.

However, other causal factors regarding the willingness to change and adopt new solutions are the mindset and attitudes of residents: strong individualism in inland Naxos and, on the contrary, a spirit of cooperation in the small community of Iraklia.

Overall, the results concerning the current situation recorded during the field study showed that:

- Travel and transport are an important part of the daily life of rural people to which considerable amounts of personal time are often devoted due to long distances to get access to services/work/extracurricular activities.
- Public transport services are poor or unavailable for remote areas (as in the case of Iraklia) and there is no basic infrastructure for the disabled and elderly.
- Private cars are the most popular mode of transport on Naxos, especially in the mountain areas.
- The only official on-demand service on the island is a taxi service, but the local people do not tend to use it very often, as it is too expensive for them.
- There is a strong dependency on the supply centre, i.e. Hora (on Naxos). Sea connections in the whole Small Cyclades are also an issue.
- Informal free-of-charge car-sharing is very popular covering people's basic needs, usually operated by neighbours or relatives.
- Cycling and walking are not popular transport modes due to the mountainous topography of the place, the lack of appropriate infrastructure and safety concerns (e.g. dangerous driving, no cycling culture). Motorbikes are used much more in comparison.

As focus groups stated, travel outside Naxos is necessary for all the residents to reach services that are not available on the island. This results in the residents being fully dependent on the maritime transport schedules and weather conditions. Ferry boats are quite frequent in the summer, but ticket prices are considered high by the residents. Flying is becoming more and more popular for travel outside Naxos (e.g. to Athens) due to the denser schedule and lower costs (sometimes even lower than ferry boats). Locals also use sea taxis to reach nearby islands to visit friends, to deal with legal issues (on the island of Syros, the capital of the Cyclades) as well as for health reasons (to reach a bigger hospital). There are three medical boats for urgent health issues, but sometimes there is no driver. In some remote areas, there are serious safety issues with small ports that cannot be used, e.g. due to bad weather conditions.

Overall, the link between the economic situation of the islanders and transport disadvantages is also relevant to their ability to support their mobility via other means. Transport conditions are difficult. The location of residents influences the volume and quality of their mobility patterns and, depending on what is offered by their administration, transport and travel can often be a drain on the energy and resources of the rural poor and remote residents. On top of this, the end of the tourist season also means reduced trips for residents, thus making it difficult for non-drivers to be transported and mainly the less advantaged and low-income residents suffer the most.

Note

1 Please see the introduction to this volume regarding the contextualisation and methodology of this research.

References

CIVITAS SATELLITE. 2019. "CIVINET Greece-Cyprus." https://civitas.eu/civinet/civinet-greece-cyprus. Accessed 27 June 2019.

Eurostat. 2011. "Population and housing census 2011, South Aegean (Greece)." https://ec.europa.eu/eurostat/web/population/overview. Accessed 25 June 2019.

Eurostat. 2018. "Hellenic Statistical Authority 2011a, 2011b, 2011c." https://www.statistics.gr/en/2011-census-pop-hous. Accessed 5 April 2020.

Institute of the Hellenic Tourism Business Association (INSTA). 2018. "The contribution of tourism to the Greek economy in 2018." http://www.insete.gr/Portals/0/meletes-INSETE/01/2019/2019_SymvolhTourismou-2018.pdf. Accessed 27 June 2019.

Kazantzakis, Nikos, and Peter Bien. 1965. *Report to Greco.* New York: Simon and Schuster.

Naxos.net travel guide. "The villages of Naxos." https://www.naxos.net/villages. Accessed 27 June 2019.

Wikivoyage. 2012. *Karte der Kykladeninsel Naxos.* Author: Bgabel auf q373 shared. https://de.wikivoyage.org/wiki/Datei:GR-Naxos-map.svg. Accessed 10 June 2020.

16 Unmet needs

Exploring mobility poverty in Buzău, Romania

Razvan Andrei Gheorghiu and Valentin Iordache

Abstract

This chapter presents the results of the fieldwork carried out in the city of Buzău, Romania, through interviews with stakeholders and focus groups with end users. The emphasis was on different vulnerable individuals, focusing on children (high school pupils), the elderly, people with visual impairments and those on low income (Roma people), all living in urban or peri-urban areas of the city. Within the city, a deprived area with poorer people and less comprehensive public transport was selected (the area of Simileasca). The main result was that, while policy makers have a weak understanding of mobility poverty, focus group participants expressed interest in innovative mobility solutions, well beyond traditional public transport systems.

Contextualisation

Buzău city is the capital of Buzău county, located in the south-eastern part of Romania near the central area of the county on the right bank of the Buzău River, covering around 82 square kilometres. On 1 January 2017, the municipality had a population of 134,552 inhabitants (National Institute of Statistics 2020). A downward trend can be observed: in 2011, Buzău had 140,875 inhabitants, but 148,839 in 2002 (National Institute of Statistics 2020). According to the Romanian National Institute of Statistics, there has been an increase in the number of elderly persons between 2011 and 2017, while the number of children and younger adults is steadily decreasing (National Institute of Statistics 2020).

The distance to Bucharest is about 100 km and the city is located at the junction of three main roads (Figure 16.1). Buzău is also an important railway node for both freight and passenger transport and has links with all areas of the country. Due to these features, it attracts transit passengers and commuters. The city has three bus stations that operate inter-urban transport. From these bus stations, private transport companies operate regular inter-urban services to other cities and municipalities in the area. With a radius of approximately 3.5 kilometres, local public transport services are currently offered on 20 main bus routes and on 10 secondary routes, all

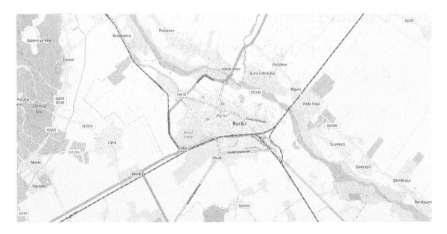

Figure 16.1 Map of the city of Buzău and surrounding areas.
Source: © The contributors of OpenStreetMap, openstreetmap.org.

provided by TRANS-BUS SA Buzău (Trans-Bus Company 2020). Public transport stations have different levels of capability and one big issue is that dynamic, real-time information regarding the bus schedule is not available. The interconnection between urban and interurban traffic is not well secured and there is a lack of intermodal transfer points. On the positive side, all the buses have facilities for persons with a disability (low floors, ramps and dedicated spaces for wheelchairs).

Buzău's modal split is indeed representative of the city's mobility: 30% private cars, 19% public transport, 1% bicycle trips and 50% pedestrian trips (Buzău City Hall 2016). The high share of pedestrian trips is mainly due to the fact that many people consider public transport inadequate: walking is sometimes faster than having to wait a long time for the bus to arrive, especially considering that the city is reasonably small. Additionally, some people cannot afford to use public transport, especially if there is no kiosk nearby and they have to buy a ticket from the driver (which costs more than those purchased at the, rarely available, kiosks).

The fieldwork, conducted in 2018 and 2019, focused on a deprived area in the western part of Buzău, consisting essentially of two districts: Nicolae Titulescu and Simileasca.[1] These areas are financially disadvantaged with many unemployed people, retired persons and several who went abroad and left their children with close relatives.

Based on a survey conducted in the study area, the main indicators for the district of Simileasca (Buzău City Hall 2017) are:

- 7.5% are people with disabilities or chronic diseases;
- 68% are unemployed people aged between 15 and 64 (compared with 0.7% for the whole city);

- 61% of the population in the area only graduated high school;
- 80% of the population in the area belongs to the Roma minority; and
- 35% are children aged between 0 and 17.

There is also a growing number of elderly people and overcrowded houses (with less than 14 m²/person) are an additional problem.

Policy makers

The municipality of Buzău is the decision-making body and it also defines the transport strategy and planning. TRANS-BUS SA Buzău is the public transport operator, which is an autonomous entity, although most of its important decisions need the municipality's authorisation. According to the municipality of Buzău, there are no significant studies and surveys covering the urban problems related to mobility poverty. Policy makers lack information about vulnerable groups and their problems and thus, naturally, they cannot formulate any action or solution. Additionally, in order to tackle mobility poverty, the municipality of Buzău would need qualified staff, a favourable regulatory framework and additional funding.

Both the representatives of the city hall and of the public transport operator mentioned that they are trying to maintain a good relationship with citizens. Communication with them is accomplished through different channels such as their websites, press conferences, press, via phone and Facebook. Polls are also conducted to monitor the satisfaction of travellers, which can also take place via the Trans-Bus website. Municipality representatives said that there were cases when suggestions have been formulated by users, triggering changes to the transport systems, such as including the specific needs of vulnerable groups.

The disparities in the treatment of the different social groups, which also includes public transport, often prompt resentment. For example, retired people, usually on low incomes, reported that they do not appreciate that socially assisted people receive benefits to an extent higher than their pension. Also, visually impaired people stated that they usually have weak support from the authorities to improve their material and financial situation, many of them being retired with small pensions or not working at all. On top of that, although public transport is subsidised for them, and disregarding their mobility difficulties, they must regularly go to the city hall to renew their monthly pass.

Hence, it is correct to say that the City of Buzău has indeed developed actions to improve the mobility of different categories of citizens, but these are not grouped in a unitary plan to tackle mobility barriers. It should be noted that the municipality and the local transport provider (due to a national enforcement) renewed the public transport fleet in order to provide exclusively vehicles with a low floor and special ramp for wheelchairs. Conversely, it looks like only physically impaired persons are considered by the authorities, who lack greater understanding of transport and mobility poverty as such. For visually impaired people, a lack of proper information systems (no tools to identify bus numbers, stops not being announced during the ride) and missing acoustic and tactile markings are important issues.

In recapping what was said during the interviews with policy makers and users, as well as in utilising the information available, we may conclude that local policy makers and transport providers have weak knowledge about the concept of mobility poverty, let alone solutions. We found the need for a better understanding of users' needs in order to address mobility poverty and transport issues in general. Due to the lack of a proper study on the topic, following our previous and current investigations, as anecdotal evidence we can state that in Romania the concept of mobility poverty is not well known as such and it is not thus considered in developing transport policy and implementing mobility systems. Moreover, a lack of funding to implement dedicated solutions for different vulnerable categories, poor legislation and regulations and even staff shortages at city hall level are significant barriers.

Main findings from the focus groups

Five focus group discussions took place in Buzău and, for each of them, we identified the following causes of mobility poverty:

1   For **low-income and unemployed**, the most pressing barrier to transport use is a lack of money along with remoteness from the city centre or from schools and facilities. Additionally, participants mentioned unsuitable bus service coverage (urban and interurban), big distances from bus stops, high fares and, finally, the lack of infrastructure for bicycle lanes and proper footpaths for pedestrians;

2   For the **elderly**, material deprivation is also important, associated with scattered social networks, physical impairment, infrequent and inadequate bus services, the low availability of interurban public transport and inadequate walking and cycling infrastructure;

3   **People with reduced mobility** (visually impaired) lack comfortable and accessible buses, appropriate information and warning systems. They are also affected by financial issues, unemployment and dependence on relatives' chauffeuring. They stated that incorrect car parking creates problems for their ability to walk around the city, requesting a stricter application of correct parking conduct. Tactile markings on the pavement for pedestrians are only present at public transport stops (but not in their vicinity). At junctions and intersections, and at traffic lights, such facilities for visually-impaired people have not yet been developed in all parts of Buzău city;

4   For **children**, the main causes of transport limitations are low bus frequency, the lack of an appropriate information system, limited accessibility to regular fares (sometimes fares are too high), limited subsidies,

infrastructure which is not suitable for cycling, including safe parking solutions. Teenagers do not hold a driving licence; and

5 **People living in remote areas** are affected by the lack of or inadequate public transport options and limited access to purchase public transport tickets. However, we also noticed complaints about the cost of fuel and car maintenance, a lack of basic infrastructures (roads, footpaths, cycle lanes) and a low degree of road safety.

Although the unemployment rate is low in Buzău, for many of the participants finding a steady, suitable or accessible job is an important issue. People who are visually impaired, of Roma ethnicity or those in the lower social strata are in many cases unemployed. Employers often avoid offering jobs to these social groups, but this situation is also the outcome of there being few employment opportunities in the city and a lower level of pay than the national average (National Institute of Statistics 2019).

While the material situation varies from one social group to another, very few of the participants have a stable job or a good pension. For students, the situation in their families is mainly good, not marked by either a shortage or abundance of wealth, but they are dependent on their parents. The Roma people are most affected by a lack of money as they are usually unemployed, some of them making their living doing different activities they find each day.

> Informal unemployment is common – people have income, they work abroad and send money home, or they have a small business in the market (without paying taxes) and almost 70% of their income is not declared.
>
> Minorities and low-income NGO representative

The elderly and visually impaired people spend their pensions or modest income wisely, taking care of priorities first and, sometimes, paying the rent can be a challenge. It is also common for elderly people to receive little support from their families, mainly due to the migration of relatives abroad. Some of them have to work to earn additional money to their pension (even as taxi drivers).

> We're too poor to spend thoughtlessly.
>
> Elderly NGO representative

All the participants are affected to varying degrees by difficulties in managing their daily travel using an organised transport system. Their needs change, sometimes daily, and, from this point of view, traditional public transport is weak in meeting these demands (and there are no alternatives to public transport). When material poverty and the absence of transport solutions come together, many participants face severe forms of mobility poverty. When prompted to discuss support from other persons, participants interviewed declared they ask relatives or neighbours for some help. In many cases, people decide to reduce their mobility considering the lack of proper solutions or support. Sometimes, they give up travelling completely or switch to walking if they are physically able to do so.

> Walking is best.
>
> High school student
>
> Elderly participant

Public transport and its issues

In Buzău, many noticeable improvements have been implemented in public transportation during the past few years. The buses are well used and the focus groups' participants declared that the travel conditions are good: buses are clean and spacious and some of them have air conditioning. The public authorities also offer different types of subsidies for all disadvantaged social groups mentioned in this study. For example, high school students and retired persons benefit from free passes, although this is applicable to one bus line only. This can cover the very basic need of going to school or to the market, but it may be that any other form of mobility (like participation in extra-curriculum courses or recreational activities) is a kind of luxury. Such a situation makes children feel socially disadvantaged, as they have many different needs according to their age and social status that would necessitate them travelling more often or in different ways. Again, behind such a decision, we can trace a narrow vision of mobility needs that is reduced to basic needs, with little attention to the users' perspective.

> Our free pass should not be limited to only one bus route. For example, I have to use one route in winter and another one in summer, which would not be accessible to me when there is snow or ice, or if it is dark outside, because I have to pass through a less travelled tunnel under the railway.
>
> High school student

Elderly people also have the option of purchasing (at reduced cost) a subscription for the entire city or receiving an additional 20 free tickets each month. Unemployed and visually impaired people are also encouraged to use public transport by being given a free monthly subscription on one bus line only. That being said, for any category, buying extra tickets is often too expensive. Students said that they need to use different bus routes and they must pay extra fees; people with disabilities must go to the City Hall every month to obtain their transport pass; people on low income often have irregular mobility patterns, as they do not have continuous work, which implies changing routes and transport needs every day; the elderly must take regular trips for bureaucratic reasons to their pensions scheme authority, which is located at the edge of the city. To answer these specific needs, a wider coverage of free access or increased subsidies would be a plus.

Besides that, accessibility to standard fare tickets is also problematic as the purchase may be made either in some specific kiosks, which are rare to find, especially in peri-urban areas, or on board where purchases carry an additional 50% cost, which may be an issue for many users. As reported in the focus groups, users then have the dilemma of seeking an official kiosk/shop (a very time-consuming activity), paying extra on board (which they cannot afford) or buying more tickets for future trips in advance (which can unbalance their weekly budget). For people living in remote areas (where selling points are very rare), this problem is more serious, pushing them to buy on board, pay extra and eventually leading to challenging situations. Often, they refuse to pay on board, sometimes trying to provoke and intimidate the bus personnel or the driver. Free public transport or easier access to normal price purchasing could easily resolve this situation.

Commuters pointed out that Buzău, due to its location, has good connections with all the major cities in the area, although it currently lacks proper connections to small surrounding villages (15 villages with about 84,000 inhabitants, with distances of between 3 and 15 kilometres from Buzău city centre) (National Institute of Statistics 2020). While users are generally pleased with the new buses, many of the participants said that the public transport service is not properly distributed in the city, questioning frequency and timetables; indeed, peri-urban areas are less well served, making it difficult for participants who live there to access it.

The children complained about the fact that the timetable is not always respected, especially at the end of the schedule, reporting that it ends too early in the evening, forcing them to use other transport facilities, which are usually more expensive.

This leads to considering the need for more frequent inter-urban transport in order to avoid crowded services as currently experienced, a situation which is affecting many social groups. Low-income persons stated that they want to benefit from the opportunity to find proper jobs in nearby cities and the children living in the cities near Buzău want to benefit from subsidies to be able to attend school, or other activities, and at the end of these activities to go back home at a convenient time.

> It does not seem right for fellow students who do not live in the city to pay for the trip to their school.
>
> High school student

Such a situation could be improved if the public transport operator integrated new lines into the existing network, or more simply, extended the current network[2].

Beyond public transport

In Buzău, the modal split reveals a low usage of private cars. Being a small city, in many cases walking is the apt choice (and indeed half of urban trips are made by walking) (Buzău City Hall 2016). Cars are indeed considered as very comfortable and fast, however, expensive. Those who own a car face the challenges of high tax and fuel expenses, so, naturally, for long distances, ride-sharing is seen as an option. When car-pooling, drivers are usually able to find a neighbour/friend to take them along for longer journeys, the fuel cost being divided among all the travellers. For short distances, apart from walking, non-motorised scooters are considered only in a few cases.

> Question from the interviewer: "Which mode of transport do you see yourself using in two years?"
>
> Answer from an elderly person: "A non-motorised scooter. An old woman riding a scooter."

Cycling can offer a lot of potential: it is NOT perceived as a "poor" system, but it is currently not a real option, mainly due to poor quality infrastructure and a lack of dedicated (secured) lanes. The poor road quality encourages car drivers to go around holes in the asphalt, zigzagging and thus creating a more unsafe environment for cyclists. This, naturally, often pushes cyclists to use the footpaths, making walking unsafe too.

> I do not agree that cyclists should use the footpaths. Both children and adults. We're dealing with them. It's not right because they run us over. They hit a lady last week. I'm afraid to walk on the footpath.
>
> Elderly participant

Also, children asked to have a secure and monitored parking space to be able to come to school by bicycle. The school has created a limited number of bicycle parking spaces, but definitely insufficient for the potential users. In the future, this should also change, as Buzău municipality is implementing projects that will redesign the main streets to accommodate bike lanes and, in addition, a bike-sharing system will be put in place to further stimulate the usage of alternative (non-motorised) means of transport in the city.

The use of ICT solutions could be useful in reducing mobility poverty, but assistance by apps or other ICT systems is mainly limited to planning travel routes when driving a private car. The use of smart devices is high among children, but they claim a lack of access to information about the public transport service. The elderly hardly use ICT solutions, very few of them accessing the Internet daily. Some of the visually impaired people use smartphones that have voice synthesis, applications related to public transport or software that may recognise objects, like tickets or different types of banknotes.

Based on the results from the focus groups, we may summarise that the root causes of mobility poverty in Buzău include:

- Low income and (too) high cost for transport (public transport, fuel, taxis);
- Remoteness from the city centre, jobs, schools and pensions scheme authority;
- Inadequate public transport coverage of the area;
- Lack of appropriate and accessible information on the (public) transport system;
- No driving licence, inability to drive, no access to a car;
- No support from family, mainly due to the migration of the workforce abroad; and
- Lack of proper infrastructure for safe walking and cycling.

Conclusions (and solutions)

In our analysis of mobility patterns in Buzău, we think there is no coherent and far-reaching understanding of mobility poverty. Beyond the lack of information, we find that an overarching framing of the problem is missing, too often reduced to offering bus access to wheelchair users. A wider understanding of disadvantaged people's needs has not yet been grasped by the local authorities or planners when designing the transport system. There is a necessity for more announcement systems (acoustic) at bus stops, more parking spaces for persons with disabilities and law enforcement for people who park private cars where they are not allowed to do so, preventing the normal movement of visually-impaired persons.

It is of course true that different groups have specific and divergent demands, but still we encountered several basic flaws, which are heavily hampering the mobility opportunity for many social groups. As mentioned above, the public transport service is not properly distributed within the city limits so it is difficult for people living in peri-urban areas to access it. Single tickets for the bus should be cheaper (with easy purchase) or free subscription should be extended to the whole network; there is a clear demand for more frequent and more reliable, on-time, interurban public transport services, with proper information provided to travellers. Currently, non-motorised means of transport are not encouraged either. The pedestrian routes in some parts of the city are of poor quality and cycling is currently not a viable option, although it is not seen as a "poor" system.

According to municipality representatives, there are, however, many projects envisaged to be implemented in the future:

- Modernising and improving road infrastructure, including two major boulevards in the city centre; implementing dedicated lanes for public transport (ensuring connections to the north, west and south);

- Implementing an adaptive traffic management system for the main axes of the city prioritising public transport;
- Development of an intermodal hub in the city in order to facilitate peri-urban–urban interchanges;
- Implementation of a new parking policy which aims to discourage private vehicle access in the city centre;
- Development of two park-and-ride systems;
- Expansion and modernisation of pedestrian areas; and
- Implementation of bicycle routes and a bike-sharing system.

All of the above should improve the mobility of these citizens, meeting many of the requirements formulated during the focus groups.

Notes

1 Please see the introduction to this volume regarding the contextualisation and methodology of this research.
2 The issues of network extension and ticketing should, however, be solved soon. The municipality has a project that will design new bus lines connecting all Buzău neighbourhoods, along with the implementation of an electronic ticketing system that will gather information about service usage, providing the metropolitan authority with suggestions about the future planning of the bus schedules.

References

Buzău City Hall. 2016. "The sustainable urban mobility plan of Buzău municipality 2016–2030: Interim report." https://primariabuzau.ro/wp-content/uploads/2016/10/PMUD-Buzau-web.pdf. Accessed 6 January 2020.

Buzău City Hall. 2017. "Strategia de dezvoltare locala a municipiului Buzău. Etapa a II-a a mecanismului DLRC" ["Buzău local development strategy – second stage of the DLRC mechanism"]. Buzău: City of Buzău.

National Institute of Statistics. 2019. "Monthly statistical bulletin of Buzău County". Issue 9. https://www.buzau.insse.ro/wp-content/uploads/2019/12/BSL0919.pdf. Accessed 6 January 2020.

National Institute of Statistics – ROMANIA. 2020. http://statistici.insse.ro:8077/tempo-online. Accessed 6 January 2020.

Trans-Bus Company. http://www.transbusbuzau.ro/. Accessed 6 January 2020.

17 Towards an understanding of the social meanings of mobility

The case of Esslingen, Germany

Tobias Kuttler

Abstract

This chapter focuses on a German case study, in Esslingen, southern Germany, investigating the mobility poverty of elderly people and refugees. In an economically dynamic region such as southern Germany, with a generally highly mobile population, impediments to a person's mobility can limit her or his opportunities and social participation. Although these groups have very different characteristics and mobility needs, both face similar challenges as regards mobility. Furthermore, this study region is the focus of attention here due to its high number of initiatives involving a very specific solution: "citizens' buses".

Contextualisation

The case study region is the district of Esslingen, located in south-west Germany, in the state of Baden-Württemberg.[1] The region is located east of the state capital Stuttgart and is part of the metropolitan region of Stuttgart.

Similar to the case study region of Luxembourg, the Stuttgart metropolitan region has a well-developed transport network, both in terms of road and rail transportation. Public transport is widely and frequently available even in smaller villages in the region. In addition, the Stuttgart metropolitan region is one of the economically most dynamic areas in Germany. Compared to other regions in Europe and Germany, the Stuttgart metropolitan region is wealthy. Baden-Württemberg has one of the highest GDP and disposable income per capita in Germany and the second lowest unemployment rate (Statistische Ämter des Bundes und der Länder 2019). Unemployment is low in the study region (Statistisches Landesamt Baden-Württemberg 2018e, 2017).

In order to identify the "blind spots" of the mobility system in the study region, this field study engages with two social groups whose basic needs and well-being are directly linked to their level of mobility: elderly people and refugees. Although, generally speaking, these groups have very different characteristics and mobility needs, they face similar challenges regarding mobility, as the following explanation will show.

Spatial, demographic and social characteristics of the study region

The study region is a sprawling, polycentric area characterised by scattered settlements alternating with industrial estates. The study region is classified as peri-urban with mixed urban (small- and medium-sized towns with historic cores and new suburban-style districts) and rural, low-density settlements.[2] Although the villages in the case study region have a rural appearance, most inhabitants are employed in the industrial and service sectors. Major employers in the car industry are located in the suburbs around Stuttgart and other major employment centres are Tübingen, Nürtingen and Reutlingen. Hence, outward commuting to Stuttgart and other industrial centres is common in the villages.

The population of the district of Esslingen grew at a rate of 3.5% between 2007 and 2017 (Statistisches Landesamt Baden-Württemberg 2018f). The study region has a slightly higher share of elderly people compared to the state average. The share of elderly people is generally lower in the city of Stuttgart and its vicinity; it increases with distance from the state capital (Statistisches Landesamt Baden-Württemberg 2018c, 2018d).

The share of population with a migrant background is higher in this region than the average for Baden-Württemberg (16.9% in 2017). Migrants mostly live in the larger cities and towns of the region, where the share of the population is between 17% and 23% (in 2017) (Statistisches Landesamt Baden-Württemberg 2018a, 2018b). A high share of refugees is accommodated in the case study region. In Germany, asylum seekers and refugees are assigned to a federal state according to a quota proportional to the total population of each state (Landratsamt Esslingen 2015). In 2015 and 2016, the district reached a peak in the number of refugees living in accommodation provided by the state.

For what is known as their "initial accommodation", refugees are accommodated on housing estates and flats provided by the district administration. A common challenge for refugees is the search for accommodation in private or municipality-provided dwellings after initial accommodation is phased out after two years, due to the tight rental market.

The focus group sessions with elderly people were conducted in **Filderstadt, Aichwald and Frickenhausen**, all in the district of Esslingen (Figure 17.1). Filderstadt and Aichwald are both located in the vicinity of Stuttgart and the city of Esslingen, while Frickenhausen is located at a distance of roughly 30 km from Stuttgart. Due to high population growth between 1950 and 1980, when the number of residents doubled or tripled, these municipalities now have a suburban character with historical cores. Commuting to Stuttgart and Esslingen and other centres of employment is very common. Aichwald is also characterised by a high share of elderly people (Statistisches Landesamt Baden-Württemberg 2018c).

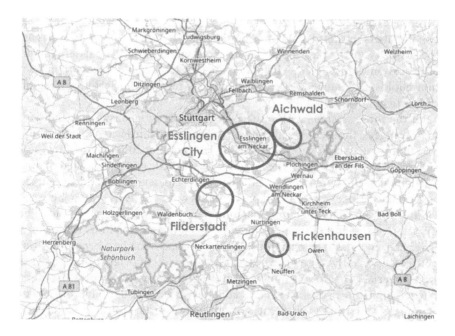

Figure 17.1 Map of the study region of Esslingen/Neckar.
Source: © The contributors of OpenStreetMap, openstreetmap.org.

Three focus groups with refugees were conducted in the city of Esslingen. **Esslingen** is a highly urbanised area in the Neckar valley south-east of Stuttgart with a high share of migrants (21.4%) (Statistisches Landesamt Baden-Württemberg 2018a). After Munich and Frankfurt, the Stuttgart metropolitan region is the region with the highest rental prices in Germany. The city of Esslingen is among the 50 German cities with the highest rental prices for newly rented apartments and will soon be at the level of the city of Stuttgart (Neuhöfer 2017). This tense housing situation has implications for people's living and mobility situation; many people decide to move farther away from the employment centres in Stuttgart, thereby increasing their commuting distances. However this often involves commuting by car.

Focus group characteristics

The three sessions with elderly people were characterised by a mix of ages. The sessions in Aichwald and Filderstadt were conducted in late summer 2018 and the session in Frickenhausen in early summer 2019. Most participants were, however, between the ages of 60 and 75, while some participants were between 75 and 88. All of the younger participants were highly active in local volunteering organisations, particularly in citizens' buses or other ride-sharing organisations. The older ones had previously been active members in their local communities. In all three focus group sessions, the majority of the participants were men, while only two participants in each session were women. All participants lived independently in their own households, in detached or terraced houses typical of the peri-urban region of Stuttgart.

Of the three sessions with refugees, the first session was conducted with male refugees, the second with women only and the third was a mixed group. The first two sessions were conducted in late summer 2018 and the third session in early summer 2019. The session with the male group consisted of nine participants, the female group consisted of 11 participants and the mixed session had 12 participants. All participants were relatively young, between 20 and 35 years old and three persons were more than 40 years old. The duration of stay in Esslingen differed between the participants: while many had been living in Esslingen for less than a year, some had been in Esslingen for between three and five years. The countries of origin of the refugees were Nigeria, Cameroon, Gambia, Iran, Afghanistan and Syria. Most of the participants in the sessions lived in collective accommodation in the city of Esslingen, either in the city centre or on the outskirts. However, some lived in the adjacent village in the hills with rural character.

Furthermore, eight expert interviews were conducted with representatives from the state ministry of transport, the regional transport department, community organisations and citizens' bus organisations.

Transport characteristics in the study region

Both road and public transport systems are very well developed in the region. Public transport coverage is comprehensive, especially in the Neckar/ Fils valleys of the region. The city of Esslingen and several other municipalities are served by Stuttgart suburban rail services and other regional trains, with journey times of half an hour or often less. The region is also served by long-distance trains.

Despite a well-developed public transport system, mobility in the greater Stuttgart region is car-based.[3] The public transportation system is concentrated on Stuttgart and the need to commute to the city. The railway routes lead radially from or to Stuttgart in the valleys surrounding the city. This also means that municipalities in the hilly areas around Stuttgart are less well connected.

Another challenge is that, outside the valleys, bus services are less frequent to major centres. In some villages, public transport is mostly oriented towards the requirements of school pupils, with low service levels in the evenings and during school holidays. Furthermore, mobility within municipalities – different districts or scattered parts of municipalities – is a challenge. Only in larger towns, such as in the city of Esslingen, there is extensive coverage of public transportation within the limits of the municipality.

Strategies and initiatives of the state of Baden-Württemberg

The state of Baden-Württemberg outlines mobility inclusion as one of the key goals in the strategy of the transport ministry "Sustainable Mobility – for all" (Nachhaltige Mobilität – Für Alle) (Ministerium für Verkehr Baden-Württemberg 2015). It recognises that especially children in urban areas and elderly people in rural areas are limited in their mobility. Besides improving public transportation services in both urban and rural areas, better walking and cycling infrastructure is stated as a key mechanism to improve mobility and accessibility for these groups. Furthermore, innovative mobility concepts and demand-responsive transportation should be implemented to improve the situation of vulnerable groups. Another emphasis is put on accessibility in rural areas. In order to tackle the challenges in a cross-sectoral approach, an inter-ministerial working group has been formed. The current range of projects in the state of Baden-Württemberg focuses on the improvement of accessibility by running pilot projects using electric mobility, car-sharing and on-demand mobility. An important public transport service in rural areas, especially at weekends and late in the-evening, is the "Ruftaxi", small buses that need to be booked (at least 20 minutes) in advance.

Several reforms of the ticketing system have been implemented recently. Since December 2018, the "BW-Tarif" allows travel within Baden-Württemberg at a standard rate. While this new tariff system simplifies travel and ticket purchasing for the end user, it also reduces ticket costs.

In April 2019, the tariff zone system of the Stuttgart regional transport authority was restructured. Due to a reduction in the number of fare zones, ticket prices have been reduced (Ministerium für Verkehr Baden-Württemberg 2018).

General mobility poverty in the study region

Due to the economic characteristics of the region, social disadvantage and social exclusion due to material poverty are rarely experienced. Furthermore, high motorisation rates and comprehensive public transport coverage in the region result in high accessibility across the whole study region. Hence, the risk of any individual experiencing mobility poverty is relatively low compared to other study regions presented in this volume.

Due to this low risk of experiencing mobility poverty, the consequence is that mobility poverty and the mobility needs of vulnerable groups are not a priority for most transport authorities and administrations in the region. In interviews with representatives of different authorities in the region, it became clear that they understand their responsibility in providing mass transport and not mobility solutions for vulnerable groups. The special mobility needs of social groups are perceived as "individual transport" and hence not the responsibility of public transportation authorities. Regarding vulnerable groups with low financial means, the responsibility is considered to lie with the social departments of municipalities and districts. Such an approach is in line with the perception that public transport should be oriented towards meeting everyday basic needs, most importantly commuting for work or education. Such an approach also determines the investment and subsidy strategy of public transport in the region: once a basic service is established in all parts of the region, further investment and subsidies are only allocated when comparatively little effort can lead to a substantial increase in ridership.

The focus on basic and mass transport needs by public authorities has two major implications for vulnerable groups in the region, while the spatial dynamics add a further challenge:

- **Disadvantages due to high cost of public transport:** first, vulnerable individuals with financial constraints may face mobility poverty when trying to meet basic needs. The ticket costs in the Stuttgart metropolitan region are high compared with other areas in Germany: hence for low-income and unemployed people, the costs can be a major barrier to mobility. A common mechanism in Germany to improve the mobility of low-income individuals is a monthly "social ticket" that is available to low-income individuals at a heavily reduced price. Such offers are made by each city and district individually for registered citizens. In the district of Esslingen, such a ticket is not offered and is not planned either. Hence, due to the different costs of public transportation tickets, the level of mobility for low-income individuals may be different according to where they live.
- **Focus on mobility needs of majority population:** when the public transport offer strongly focuses on serving the needs of the majority population as it does in the region of Stuttgart, vulnerable individuals may experience mobility poverty because the transport offer does not fit their circumstances. Generally, the public transport coverage is best during commuting hours on weekdays. Hence, those who work late at night and in the early mornings or at weekends may experience mobility barriers. Another barrier exists for people with reduced mobility. Unlike in other German regions, special mobility services (door-to-door) for disabled people do not exist.
- **Increasing spatial disparities:** due to high rental prices in the main cities of the study region, lower income groups have to move further away from their places of work. For the future, this could mean that the population in rural areas is increasingly either old age or on low income. Due to limited public transportation, the mobility of low-income groups is car-based, leading to incidences of forced car ownership. Interviews revealed that these interrelations and future challenges are well understood and tackled by the transport ministry of Baden-Württemberg.

However, at local and regional levels, there is a tendency to underestimate these challenges, partially because sectoral administrations do not have the capacity to tackle complex cross-sectoral tasks.

Mobility poverty of elderly people

Transport situation of elderly people

The elderly in the region heavily rely on private motorised vehicles. All participants in the focus group sessions drove their own cars except for two participants. The car was the main mode of transport, although senior tickets are available at reduced prices. The participants expressed that losing the ability to drive one's own car is often a turning point in one's life, but most of the younger elderly do not like to think about future times when they would not be able to drive any more. Instead, they insist on driving as long as possible. One participant stated:

> When you sell your car, it is perceived that one is not mobile any more. That is a huge barrier. There are many alternatives to driving a car, but there is no awareness of these alternatives among many elderly.

Due to these attitudes, other options are either not known or not seriously considered.

The distinction between mobility for basic everyday needs and mobility for social purposes is effective when analysing transport and mobility poverty in the German study region.[4] The elderly people who participated in the study mostly had mobility needs for everyday requirements in the local area such as grocery shopping, appointments with their doctor. Furthermore, the participants travelled for social purposes such as to meet relatives and friends as well as to participate in social and cultural activities. Many focus group participants reported that they visit their grown-up children in other parts of Germany and also drive there by car.

Meeting basic mobility needs

There is a distinction between basic needs and social purposes because, as already mentioned above, for the majority of the elderly, mobility poverty in the narrow sense is not an issue in this region. However, elderly people with low financial means may face challenges due to the high cost of public transport and a mobility system that centres around the working population. The discussion in the focus groups revealed that elderly people without access to a car are vulnerable to mobility poverty and social exclusion. This vulnerability is increased when elderly people are physically impaired or have a disease that limits their mobility. In these cases, even though public transport coverage in the region is extensive, public transport stations and bus stops are often too far away from homes. This can be observed in particular for the estates with lower density outside the historical cores of towns and villages. The use of taxis may be limited due to low financial means or because taxis do not serve villages outside the urban cores.

Furthermore, participants in the focus group session acknowledged that a low-income situation may be a social stigma in a generally wealthy region. Many people would not easily accept help but try to solve a crisis by themselves, as a community organiser highlights:

> I know people in our town who are really struggling to make ends meet and they never seek help. There is a certain pride in saying 'I want to manage without external help.'

Travel for social purposes and the social function of mobility

Participants in the focus group session considered themselves as financially independent and hence saw their basic mobility needs covered by either public or private transport.[5] Instead, they greatly emphasised the need to travel for social purposes and the social function of mobility itself. They emphasised that social participation is key to well-being for those who are very old and mobility plays a key role in preventing social isolation.

Many of the participants still drove their own cars, but expressed that driving a car alone contributes to social isolation. Hence, they highlighted that traveling together in groups for leisure trips on public transportation has become very popular among the elderly in the region, supported by the availability of a "senior ticket" at reduced rates. Furthermore, offering rides to neighbours or friends is a way of giving support, but it is also a low-key effort to stay in touch and socialise. Mobility is hence not only the "vehicle" to get to locations in their social life, but rather mobility itself becomes the social event.

During the focus group sessions, another social implication was observed among the participants. Mobility was one crucial element of a person's self-image. Many proudly reported long distance leisure trips or visits to family members who live far away. Several participants highlighted how much they liked driving cars over long distances. On the one hand, for those who enjoy a high standard of living in old age and have been in executive or skilled positions before retiring, this "travelness" is an expression of independence and freedom on the one hand and a means of social distinction and reputation among the elderly on the other.

Mobility poverty of refugees

Role of mobility in employment and education

The participants in the focus group sessions considered employment and education as the most important prerequisites for their well-being. Hence, mobility related to these purposes is of prime importance.

Besides walking, refugees in the study area heavily rely on public transport. None of the focus group participants owned a car. For those with direct access to trains (in the valleys of the Stuttgart region), it is easier to find opportunities for employment and education in other parts of the region. Access to jobs is more difficult if one lives in the hilly regions of the area with bus public transport. Then, cycling becomes an important means of mobility for the male refugees, as one participant explained:

> I live outside of Esslingen and I work late at night, early morning or weekend shifts. During these times, there are no buses running to the next train station so I ride my bicycle to the train station. Only then can I get to work on time.

Related to the problem of accessibility is the system of assigning accommodation to refugees by the district authorities. Accommodation can be provided in both city centres and in more remote villages. The participants expressed that this system seems like a "lottery system" to them and it can limit their opportunities to find employment or suitable education.

Impact of high ticket costs and complexity of tariff system

When taking up employment, refugees usually receive low-income jobs, but then they must also pay for their accommodation on their own. This means that their financial situation does not necessarily improve substantially once in employment. Given these constraints, public transport in the region is costly for the refugees. Also, refugees with a reduced budget rely on grocery shopping at discount shops, which are not available in all villages where they live. Hence, they have to do grocery shopping at costlier supermarkets or have additional expenses to travel to a discounter. Participants in the focus group sessions also raised concerns about the costs of travelling with children above the age of six and about the latest increase in ticket prices.

Participants mentioned that for refugees who have recently arrived in Germany it is difficult to understand the complex public transport tariff and ticketing system. The participants were very aware that improving their language skills helps when using public transport. They valued the efforts made by volunteers and the municipal social services that support newcomers by explaining how the public transportation system works.

Mobility of refugee women

Specific findings on the mobility problems of women refugees were obtained through the women-only focus group. Women have to travel longer distances than men, e.g. to attend a specialist doctor. The language problem turned out to be more severe for women because some of them were illiterate and had difficulties in using a ticket machine. Women also travel more often with young kids on public transport.

Although almost half of the women in the focus group sessions knew how to ride a bicycle, bike usage among women was not common in Esslingen. Many women expressed that they would be eager to learn how to ride a bicycle. One women stated:

> I would love to ride a bicycle, but I do not know anyone who could teach me how to ride.

However, the women expressed that there are constraints when riding bikes such as travelling with children.

Discrimination on public transport

Refugees from African countries reported that they face discrimination on public transport. They highlighted that especially when young African men travel in groups, they feel targeted by ticket inspectors. One participant stated:

> The inspectors do not have much time, they cannot check everyone so that's why it is happening to us refugees.

The participants reported that they are sometimes caught without valid tickets and they are severely fined even in situations when they are travelling with a ticket, but the ticket is wrong or not valid any more. The participants raised the concern that there is not enough awareness of the difficulties they face when not acquainted with the public transport system.

Role of community-based, volunteer mobility services: citizens' buses

In more and more municipalities in Baden-Württemberg, forms of community-based transport are being implemented. The most common form is the citizens' bus (Bürgerbus), a public transport service which relies on unpaid volunteers for most or all tasks, in particular to drive the vehicles. The majority of the funding of a citizens' bus is raised locally, e.g. via donations from local businesses. The vehicles used are usually minibuses or large passenger cars. While citizens' buses usually run on fixed schedules and routes, there are also flexible services that have the nature of an on-demand service.

Community transport solutions were first implemented in Germany in 1985. Since then, the number of services has increased to more than 250. In Baden-Württemberg to date, 84 citizens' buses are running and an additional 12 on-demand services. In addition, there are numerous voluntary services with restricted public usage such as ride-sharing and car-pooling services for elderly people. In the district of Esslingen, citizens' buses have been in operation since 2003. Citizens' bus initiatives are embedded in their local communities and very well connected to their local administrations and economy.

Citizens' buses are often launched and operated by local senior citizens' organisations and they are mostly frequented by elderly people. However, they are open to the general public. Refugees and other social groups with disadvantages, such as low-income groups, are not usually riders on citizens' buses and are not specifically targeted by citizens' bus initiatives. However, there are exceptions, such as the initiative in Aichwald, which specifically targets low-income persons via a special programme.

The attractiveness of citizens' buses for elderly persons can be understood when considering the social significance of citizens' buses. Many of those who volunteer as drivers for citizens' buses are younger elderly men, those who recently retired and are searching for meaningful activities for their free time as well as social interaction. The passengers on citizens' buses, on the other hand, are often elderly people who have lost their partners and /or other close friends/relatives. At this older age, one's own physical abilities may already be reduced. While using the citizens' bus for basic daily needs, the purpose of social participation is a crucial objective of these "older" elderly. Community organisers explicitly highlight the significance of these services for social inclusion. An organiser of community transport in Filderstadt stated:

> The social isolation of elderly people is one of the core problems. Our mobility service contributes to reducing isolation in older age.

Here, the difference to conventional public transport becomes apparent. The participants in the focus groups with senior citizens reported that, while they rarely use conventional public transportation, they regularly use citizens' buses and other forms of community-based transport. They even decide to use citizens' buses for certain trips instead of their own cars, for the sake of the social interaction with the drivers and other passengers. For some participants, a ride on the citizens' bus has become a regularly scheduled "social event". The citizens' buses also strengthen the networks of the elderly in the villages and towns. A citizens' bus organiser from the town of Wendlingen stated:

> Organisers, drivers and users of the citizens' bus have become a committed community here. People identify with the citizens' bus and a sense of cohesion has developed.

Also, volunteers enjoy the appreciation they receive for their work, as he adds:

> There is a lot of gratitude from the citizens and the mayor towards the organisers and drivers. In a small town, people often need this kind of public recognition for their commitment to a social cause.

Evidence from Baden-Württemberg shows that, among transportation administrations and authorities, the motivation to collaborate with and support citizens' bus associations is limited. One of the reasons is that they do not see a need for additional transport services because conventional public transport is extensive. However, as outlined, riders of citizens' buses often – implicitly or explicitly – travel for social motivations, sometimes the social motivation is even more dominant. Citizens' bus associations, hence, often complain that local and regional public transport administrations misconceive the nature of citizens' bus services.

To close this gap between public authorities and volunteer initiatives, Baden-Württemberg is one of the few states in Germany that actively supports the development of community-based mobility solutions such as citizens' buses. In order to channel the efforts of the state of Baden-Württemberg, the "Competence Centre for New Forms of Public Transport" (Kompetenzzentrum neuer ÖPNV) at the Nahverkehrsgesellschaft (Local Public Transport Agency) Baden-Württemberg (NVBW) is exploring new approaches to mobility especially in rural areas and supports implementation with counselling and additional funding.

Conclusion: beyond basic travel needs

The results from the field study with the elderly and refugees in the region of Esslingen highlight two major observations: first, even in an economically dynamic and well-equipped region, concrete and severe mobility problems may arise for members not part of the majority working population; second, the study revealed the need to recognise the social dimension of mobility when analysing mobility poverty. This has both conceptual and practical implications. The implications are conceptual because drawing attention to the social significance of mobility offers a rich pathway to understanding the phenomenon of mobility poverty in its respective context, but also recognises that mobility disadvantage is structural and systemic. However, these implications are also practical: recognising the importance of mobility for social purposes addresses the challenges and desires of a much larger part of the population.

Transport administrations and authorities do not yet give enough consideration to these social aspects of travel and the wider benefits for the well-being of socially vulnerable persons; and they do not consider these aspects as part of their mandate. However, growing state support for citizens' buses in Germany is an indicator that there is a growing awareness of the many dimensions of mobility.

Notes

1 Please see the <u>introduction</u> to this volume regarding the contextualisation and methodology of this research.

2 The population density in the case study region ranges from 70/km^2 in rural areas to 1966/km^2 in the cities (Statistisches Landesamt Baden-Württemberg 2018f).

3 Motorisation rates range from 609 vehicles per 1000 inhabitants (in Filderstadt in the vicinity of Stuttgart) to more than 850 in rural municipalities in the region (Statistisches Landesamt Baden-Württemberg 2019).

4 This differentiation does not ignore the fact that the basic needs and social needs of elderly people are often connected. As Hjorthol (2013, 1203–1206) writes, meeting the more basic needs like shopping receives wider significance for an elderly person's well-being, for example due to the personal assurance of being independent and in control of one's life, the possibility to meet friends, or just the positive feeling of being out of the house, "on the road" or among people.

5 Those older people who are materially deprived in the case study region are largely invisible, both to the general public and this research project. On the one hand, this is an almost inevitable consequence of the research approach itself. In order to gain access to individual users, these individuals were approached via associations that network among and support elderly people. Hence, only those who are well networked, are interested in research questions or are actively engaged in voluntary community work participated in the focus group sessions. These individuals are, at the same time, the younger elderly and the ones who are stable financially and health wise. Furthermore, the overall picture of the high quality of life and comparatively few challenges is reproduced by the research participants themselves who mostly report from their own experiences.

References

Hjorthol, Randi. 2013. "Transport resources, mobility and unmet transport needs in old age". In *Ageing & Society* 33, 1190–1211. https://doi.org/10.1017/S01446 86X12000517.

Landratsamt Esslingen. 2015. Flüchtlingsarbeit im Landkreis Esslingen. Esslingen am Neckar. https://www.landkreis-esslingen.de/site/LRA-Esslingen-ROOT/get/ params_E750212913/15395780/Handreichung%20Fl%C3%BCchtlingsarbeit%209-2018.pdf. Accessed 21 January 2020.

Ministerium für Verkehr Baden-Württemberg. 2015. Strategie „Nachhaltige Mobilität – Für Alle". Stuttgart. https://vm.baden-wuerttemberg.de/fileadmin/ redaktion/m-mvi/intern/Dateien/Brosch%C3%BCren/Neue_Mobilitaet_Strategie broschuere_Baden_Wuerrtemberg.pdf. Accessed 21 January 2020.

Ministerium für Verkehr Baden-Württemberg. 2018. Projects and Goals. 3rd updated edition. https://vm.baden-wuerttemberg.de/fileadmin/redaktion/m-mvi/ intern/Dateien/Brosch%C3%BCren/Zwischenbilanz_2018_EN_final.pdf. Accessed 21 January 2020.

Neuhöfer, Manfred. 2017. F+B-Wohn-Index Quartalsbericht I 2017. F+B Forschung und Beratung für Wohnen, Immobilien und Umwelt GmbH. Hamburg. http:// www.f-und-b.de/files/fb/content/Dokumente/News/F+B-Wohn-Index_2017Q1. pdf. Accessed 21 January 2020.

Statistische Ämter des Bundes und der Länder. 2019. Volkswirtschaftliche Gesamtrechnungen der Länder. https://www.statistik-bw.de/VGRdL/tbls/tab.jsp. Accessed 21 January 2020.

Statistisches Landesamt Baden-Württemberg. 2017. Interaktive Karten: Arbeitslosenquote bezogen auf alle zivilen Erwerbspersonen 2017e. https://www. statistik-bw.de/Intermaptiv/?re=kreis&ags=08116&i=07501&r=0&g=001&afk= 5&fkt=besetzung&fko=mittel. Accessed 21 January 2020.

Statistisches Landesamt Baden-Württemberg. 2018a. Interaktive Karten: Ausländeranteil 2018 (Gemeindeebene). https://www.statistik-bw.de/Intermaptiv/?re= gemeinde&ags=08111000&i=01401&r=0&g=0001&afk=5&fkt=besetzung& fko=mittel. Accessed 21 January 2020.

Statistisches Landesamt Baden-Württemberg. 2018b. Interaktive Karten: Ausländeranteil 2018 (Kreisebene). https://www.statistik-bw.de/Intermaptiv/?re=kreis& ags=08116&i=01401&r=0&g=001&afk=5&fkt=besetzung&fko=mittel. Accessed 21 January 2020.

Statistisches Landesamt Baden-Württemberg. 2018c. Interaktive Karten: Durchschnittsalter 2018 (Gemeindeebene). https://www.statistik-bw.de/Intermaptiv/?re= gemeinde&ags=08315113&i=01301&r=0&g=0001&afk=5&fkt=besetzung& fko=mittel. Accessed 21 January 2020.

Statistisches Landesamt Baden-Württemberg. 2018d. Interaktive Karten: Durchschnittsalter 2018 (Kreisebene). https://www.statistik-bw.de/Intermaptiv/?re=kr eis&ags=08116&i=01301&r=0&g=001&afk=5&fkt=besetzung&fko=mittel. Accessed 21 January 2020.

Statistisches Landesamt Baden-Württemberg. 2018e. Interaktive Karten: Bevölkerungsdichte 2018f (Gemeindeebene). https://www.statistik-bw.de/Intermaptiv/? re=gemeinde&ags=08125038&i=01202&r=0&g=0001&afk=5&fkt=besetzung& fko=mittel. Accessed 21 January 2020.

Statistisches Landesamt Baden-Württemberg. 2018f. Interaktive Karten: Bevölkerungszu- (+) / -abnahme (-) 2007 bis 2017 (Kreisebene). https://www.statistik-bw. de/Intermaptiv/?re=kreis&ags=08116&i=01204&r=0&g=001&afk=5&fkt= besetzung&fko=mittel. Accessed 21 January 2020.

Statistisches Landesamt Baden-Württemberg. 2019. Interaktive Karten: PKW 2019 (Gemeindeebene). https://www.statistik-bw.de/Intermaptiv/?re=gemeinde&ags= 08237045&i=20106&r=0&g=0001&afk=5&fkt=besetzung&fko=mittel. Accessed 21 January 2020.

18 Mobility poverty in Luxembourg

Crossing borders, real estate, vulnerable groups and migrants

Patrick Van Egmond and Joanne Wirtz

Abstract

The economic and political relevance of the city of Luxembourg to its Greater Region leads to increased mobility issues (along with urban sprawl). The main urban centre is like a magnet that pulls people and goods, but, on the other hand, pushes out low-income residents, due to the high cost of living. This chapter presents the results of fieldwork carried out in Luxembourg with a focus on the mobility poverty of migrants and persons living in rural areas, in particular cross-border areas. There is not much recognition of mobility poverty as a topic in the Greater Region. Nevertheless, a lack of efficient connections between the transport systems of neighbouring countries, the hinterland and Luxembourg is a regular topic of debate. Luxembourg is heavily investing in cross-border infrastructure and its regional bus network to improve this situation. Free public transport will reduce the cost of transportation, yet high costs of housing and further economic growth will most likely worsen the mobility situation of some vulnerable groups.

Contextualisation

Luxembourg is a small country of 2,586 square metres, with about 602,000 inhabitants (Statec 2019). However, for transport-related issues we should consider the Greater Region, which comprises Luxembourg and neighbouring parts of France, Belgium and the Saarland region of Germany. From the 1960s onwards, Luxembourg was transformed from a mining and steel production-led industrial area to an important banking and service hub. The declining heavy industry sector led to increased poverty in the former steel industry area in Luxembourg and bordering areas of Belgium and France.

Today, most of the economic activity in the Greater Region takes place in the south of Luxembourg, around the capital. Most of the jobs are concentrated in and around the city of Luxembourg. The country's transformation led to an increasing cross-border exchange of labour towards Luxembourg. Nowadays, almost 45% of the approximately 432,000-strong national workforce commutes daily between Luxembourg and neighbouring regions in Belgium, France and Germany. The north of Luxembourg can be considered rural. The area is more dispersedly populated and accounts for about 80,000 inhabitants. Accessibility to public transport both in the former steel areas and the north is less dense than in the economic centre.

Luxembourg also welcomes about 2,000–2,500 refugees annually. The Luxembourg National Reception Office (ONA) oversees the Grand Duchy of Luxembourg's reception and integration policy. The integration policy applies to all foreigners, i.e. to European Union citizens as well as to third-country nationals. ONA also organises the reception, accommodation and social supervision of applicants for international protection. All refugees are offered a free public transport pass.

Luxembourg has the highest number of vehicles per capita (669 cars /1,000 inhabitants) in Europe. About 97% of households within Luxembourg have at least one private car (Statec 2019). On average, there are about 1.4 cars per household. Most households without a private vehicle are in the city of Luxembourg and they are mostly single-person households whose age is between 18 and 24 years old, followed by those aged above 75.

The main public transport provision is organised by the Ministry of Mobility and Public Works for the regional transport (RGTR), by the city of Luxembourg (AVL) and by the communities of the south (TICE). Train transport is organised by the national rail operator (CFL). The Transport Community (Verkéiersverbond) is responsible for the ticketing and information system.

Since 1 March 2020, public transport is free for the whole country when travelling second class. While tariffs remain for those travelling in the cross-border areas, most of them benefit from this new measure during their daily commute. In addition to the main public transport offer, specific transport services are being offered for the residents of Luxembourg with specific needs. The transport offer is organised and adapted to the needs of people with reduced mobility, allowing them active participation in social life and an increase in autonomy.

Initiatives

The "CAPABS" service transports students who need differentiated education as well as disabled employees to social workplaces or centres for the physically and/or multi-handicapped. In 2016, this service carried out 686 trips daily and is organised by the Ministry of Mobility and Public Works.

The "ADAPTO" service provides transport for employees with reduced mobility to their place of work as well as occasional transport for disabled citizens. At the end of 2015, about 5,500 residents had an Adapto card (CES 2017).

The service can be used for occasional trips and may be taken up to a maximum of 15 times a month. The service is available in the territory of the Grand Duchy of Luxembourg seven days a week from 7 am to 10 pm (on Fridays and Saturdays, departures are allowed until midnight). In addition, the Adapto services have become free of cost since March 2020. A trip has to be booked long enough, at least one and up to two days, in advance. A passenger may use any of the 27 aggregated transporters.

The Bummelbus is transport on demand running in the northern part of Luxembourg and it includes 39 municipalities. The Bummelbus is a supplement to available public transport and private means of the 80,000 inhabitants in those northern municipalities. All citizens have access to this service for short-distance journeys, especially within the municipalities and neighbouring villages/towns. The service is presently run using 50 vans (in up to nine places). Part of a social reintegration initiative, the drivers are all unemployed and are offered an opportunity to requalify for the job market through a two-year maximum contract.

Focus group characteristics and definitions of vulnerable groups at risk

There is little recognition of mobility poverty as a topic in the Greater Region. Nevertheless, a lack of efficient connections between the transport systems of neighbouring countries and Luxembourg is a regular topic of debate.

In Luxembourg, among the groups most at risk are migrants and people living in remote rural and deprived areas. We should also consider cross-border workers. In the preparatory stage of the fieldwork, conducted in 2018 and 2019, it became clear that a more careful distinction between vulnerable groups needs to be developed, considering that there is a combination of multiple factors that determines the risk to mobility poverty. Naturally, not all migrants are at risk of mobility poverty, nor all persons living in remote and rural areas.[1]

Migrants

In relation to migrants, a distinction must be made between refugees who receive political shelter and mainly Europeans who migrate to Luxembourg for work purposes. Moreover, there is a distinction between migrants in highly paid employment and migrants in a less favourable economic situation. The migrants participating in the HiReach focus groups, living in urban centres, considered themselves relatively well served in terms of transport options; more problems arise for those living in remote and deprived rural areas with lower levels of income. The focus group discussions made it clear that even if the latter group had free public transport, the poor offer of transport connections hindered their integration. Importantly, many of them had no resources presently available to buy a private car.

People living in remote rural and deprived areas

There is no one single level of mobility poverty for persons living in remote rural, deprived and cross-border areas. There is a clear link with the socio-economic situation of the household. In the case of Luxembourg, the level of mobility poverty bears some relation to nationality. During the focus groups, it was reported that people travelling cross-border and dwelling beyond five kilometres from the border have higher levels of mobility poverty due to a lack of available cross-border transport services.

Comparing the results of women and men in the focus groups, it was evident that women living in remote rural and deprived areas can be identified as a vulnerable group. This relates to the specific wider socioeconomic situation of the household. When a single car is available, it is often the man who uses it. Also, safety, harassment and fear of potential aggression were listed as causes of limiting full use of the transport offer. This was even more applicable to public transport in the evenings or at night and indeed to services such as car-pooling too. "Women only" options were favourably considered.

Cross-border commuters

The focus groups overall brought together a large group of nationalities (e.g. French, Cameroon, Syrian, Iranian, Kosovan), but the focus group with cross-border commuters was more homogenous, as only French citizens participated. This reflected the constituency of the cross-border population in the Greater Region on the French side. Indeed, many different nationalities live in the country. Only 52.1% of the population is of Luxembourgish nationality, of which a high proportion has parents of another nationality. Over 16.4% of residents have Portuguese nationality, 6.6% French and 21.5% have another nationality (Statec 2019). The cross-border commuters and families are mostly the nationality of residence. Presently, families including Luxembourgish, move cross border due to the high prices of housing.

There is a large difference in terms of risk of poverty per nationality and social group within Luxembourg. The poverty risk is as high as nearly 12% for residents with Luxembourgish nationality, yet rises up to 30% for those with Portuguese heritage (Statec 2017). This seems mainly related to the difference in access to the job market as well as the time spent in the country. The share of cross-border commuters amongst those employed decreases the higher one moves up into management level.

About 5% of the population suffers from material and social poverty and 15% has an income lower than 60% of the average income (21,200 euro per year). Most people suffering from poverty are single parent families. Likewise, there is a risk for the elderly aged over 65, yet these groups are compensated with social support.

Tertiary service jobs (e.g. in shops, restaurants, day care and commerce) are filled by cross-border commuters and other jobs (e.g. cleaning, company catering and the construction sector) by lower skilled residents of which the latter is made up of a large share of residents with Portuguese nationality. There is a trend of eastern European workers entering this specific job market.

Women

This focus group specifically brought together women working in the tertiary sector who commute daily across the border. They have in general complex work patterns and related travel patterns. They leave home very early and arrive back late in the evening, spending about one hour and 15 minutes travelling each way. One of the participants also reported her experience as a holder of two jobs: one in the morning and one later in the afternoon/evening, both in Luxembourg. She leaves home early in the morning by car. She leaves her car at the first workplace. Then, in the early afternoon, she sleeps for two hours in her car and then takes public transport (a short train trip plus bus) to the second job. She leaves work at 7 pm and goes back to her car, arriving home at 8 pm. As confirmed in the focus group discussion, poorer households in particular spend up to 30% of their household income on mobility, which is far above average.

Compared to France, the advantage for cross-border workers is the higher wages in Luxembourg. The price to pay is the higher financial cost and more intrusive time budget for travelling, which is mainly based on the private car, well-functioning public transport along the main corridors and alternative forms like car-pooling.

The link between mobility poverty and housing costs

In Luxemburg Greater Region, accessibility and available forms of transport are only partly considered when searching for a place to live. Those especially at a higher risk of economic poverty are pushed away from Luxembourg City and its surroundings towards the less expensive dwelling alternatives of the Greater Region. The focus group discussions clearly pointed out that participants i) living in rural and deprived areas (but working in and around the capital of Luxembourg) as well as ii) living in cross-border areas find it difficult to live closer to their place of work due to the real estate market conditions.

Findings from the focus group sessions show that there is a clear link between housing costs and mobility. Due to the concentration of employment in the city of Luxembourg, the average advertised housing price differs greatly between Luxembourg's economic centre and its other areas, as well as cross-border areas. In 2015, the average advertised price for the purchase of an apartment in the city of Luxembourg was around 7,100 euro/m² compared with 5,200 euro/m² for the rest of the country. This is similar for houses: 5,500 euro/m² for one in the city of Luxembourg compared with 3,900 euro/m² for the rest of the country (CES 2017).

The same differences exist for the cost of land suitable for housing. The median price of land located about 30 minutes from Luxembourg-Ville is about 38% of the median value of the capital. At a driving distance of 60 minutes from the capital into the heart of the north and cross-border areas, the median price is five times lower than the median price recorded in Luxembourg City.

Mobility poverty and accessibility

In the cross-border focus group, none of the participants currently uses public transport. They did not live close to any public transport that would bring them quickly to Luxembourg. They could drive to one of those stops, yet then prefer to organise a car-pooling scheme due to the low and poorly organised frequency of the bus timetable. In the focus group on rural and deprived areas, some trips were made using public transport due to a good direct regional bus connection and the place of work. However, this trip was only made when they did not need to pick up children (i.e. somebody else, as the spouse took care of family and household duties).

We can now better understand that only 7% of the daily cross-border commuting is done by public transport, either as the sole mode for about one third of people travelling or by combining it with a private car (park and ride). This low use of public transport is also the consequence of weak accessibility and multiple destinations for the commutes (e.g. school, work, shopping, grocery shopping).

A study carried out in 2012 (Schiebel et al. 2012, 10) identified that 40% of the cross-border regional public transport stops, mainly bus stops, are close to the Luxembourg border, in other words up to 5 km from the border. About 12% are located within 5 km and up to 10 km from the border. About 27% of the stops are in a range from 10 km up to 30 km yet are mainly train stops. From a combined spatial-transport offer point of view, there is a large difference in accessibility during peak hours towards the economic centre of Luxembourg City.

On an accessibility map created by Julien Schiebel and his team (Schiebel et al. 2012), it is shown that when it comes to accessibility, Luxembourg would contract on the main car and train arterials towards Luxembourg City. Via the main highway and mainly rail-based public transport systems, locations within Luxembourg have better access to the country's economic centre, yet all trips from cross-border residences outside of these transport corridors are comparatively long. In the case of a trip with a destination other than Luxembourg City, or in the case of multiple destinations, journeys become significantly longer and accessibility is substantially reduced. While the population in the southern area benefits from relatively high access to public transport, such an offer is weak in the northern parts of the country.

There is a combination of the concentration of economic activity in the south, large differences in real estate market prices and yet also a widely shared desire to access home ownership. For Luxembourg, the rate of home ownership is 70.8% (STATEC 2014). This seems to lead to an increasing distance between home and job location. In particular, younger generations tend to move further away from the cities towards the urban periphery or rural areas, or even beyond the borders of the Grand Duchy. This creates extended urbanisation, resulting in longer journeys and an increased need for mobility, often triggering mobility poverty due to the gap between supply and demand.

Luxembourg and refugees

The Luxembourg population is growing annually with about 2% (circa 12,000 inhabitants). This is mainly through European citizens migrating to the country. As mentioned above, Luxembourg welcomes about 2,000 to 2,500 refugees annually (MAEE 2017). One of the focus groups was devoted to this social layer and its discussion showed mobility poverty as a result of dwelling choices. The reasons for refugee's mobility include seeing friends, personal appointments, education and sport. Due to current legislation, none of the participants had a regular job and they have been offered housing both in rural and remote areas of Luxembourg or in Luxembourg City. Most of them use the bus as their primary mode of transport to commute to Luxembourg city centre on a regular basis.

However, often the last buses in the evening are not suitable to allow social life after the late afternoon. The frequency of buses at the weekend is also low (i.e. one bus per hour), which results in limited and planned-in-advance mobility. Due to their financial situation, taxis are widely regarded as very expensive by refugees and therefore not considered a mobility option. Some refugees use a car for the first and last miles and then change to a train to go to the city centre, due to traffic jams; yet there are also parking problems and high parking prices in the city.

Another outcome is that, generally speaking, refugees have no problems in understanding the public transport system; younger refugees in particular are well versed in the internet and smartphone applications. It was reported that, as soon as they have the financial means and possibility of using their driving licence, they will buy a car and use it as their primary mode of transport. The car is also seen as a convenient mode to travel outside of Luxembourg and to do weekend trips.

Shopping is usually done close to home. However, evening courses for continuing education can only be reached by car, as there is no bus connection for the return journey. Most of the activities are organised in urban centres, which is why some activities (e.g. sports classes) cannot be undertaken if one lives in a remote village and the activity takes place on a Sunday. This is even true for activities in the city centre, as there is also only one bus per hour on Sunday (and none during the night) with bad connections on Saturdays.

For the refugees who live outside of the city centre, this leads in some cases to social exclusion, yet the risks may be higher than for the rest of the population. This mainly relates to social activities which foster integration into society. Likewise, when they obtain legal status to access employment, success on the labour market is restricted by low transport accessibility.

Conclusions

Considering the economic and political relevance of Luxembourg City, we can conclude that this urban centre is a magnet that pulls people and goods, but, on the other hand, it pushes away workers with lower wages due to the high cost of living. This creates mobility tensions with the rural parts of the Duchy and triggers increased cross-border traffic. Almost 45% of the approximately 432,000-strong national workforce commutes daily between Luxembourg and neighbouring regions in Belgium, France and Germany. Looking beyond the country's borders, apart from some medium-sized cities (Trier, Arlon, Thionville and Metz), the Luxembourg Greater Region can be considered rural or a dispersed form of urban sprawl. The choice of living in a specific part of the northern or south-eastern part of the country, or even in the Great Region, is the result of multiple factors and transport is one of the variables in such a choice. This does not apply to the majority of refugees, who have little freedom of choice regarding their settlement.

To sum up for the vulnerable groups investigated here, the causes of mobility poverty are a mix of the socio-economic situation and specific social motivations (family ties, personal preferences). Cross-border workers gain income by working in the country of Luxembourg, yet lose significantly in terms of their time budget and travel costs. The current public transport offer leads in most cases to a strong car dependency. Car-pooling or increased inter-modality could be a solution. As cross-border households are close to public transport infrastructure or are well off due to their socio-economic situation, they are not all considered mobility poor. Nevertheless, the increased use of private mobility (i.e. a car) impacts their household budget.

The focus group discussions also showed a correlation between a lack of accessibility to transportation and an increased risk of social exclusion. Mobility poverty prevents people from taking an active part in society. This was specifically the case for workers living in remote rural areas, with daily (cross-border) commuting to work, school and social activities. When depending on public transport, social activities are more restricted. Another interesting point is that the time spent in cars is a significant proportion of the day, thus reducing time spent on other social activities.

Note

1 Please see the <u>introduction</u> to this volume regarding the contextualisation and methodology of this research.

References

Conseil Economique et Social de Luxembourg (CES). 2017. "Perspectives économiques sectorielles à moyen et long termes dans une optique de durabilité.". *Transport, Mobilité et Logistique.* https://ces.public.lu/dam-assets/fr/avis/politique-generale/transport1.pdf. Accessed 6 April 2020.

Ministère des Affaires Etrangères et Européenne (MAEE), Direction de l'Immigration. 2017. "Statistiques concernant la protection internationale au Grand-Duché de Luxembourg Mois de mars 2017." https://maee.gouvernement.lu/content/dam/gouv_maee/directions/d8/publications/statistiques-en-mati%C3%A8re-d-asyle/2019/Mars-2019-Statistiques-protection-internationale.pdf. Accessed 6 April 2020.

Schiebel, Julien, Sylvain Klein, and Samuel Carpentier-Postel. 2012. Simulation de l'accessibilité en transport en commun transfrontalier vers le Luxembourg. Association de Science Régionale De Langue Française, 49e colloque de l'Association de Science Régionale De Langue Française, "Industrie, villes et régions dans une économie mondialisée., UMR ThéMA; ASRDLF, Jul 2012, Belfort, France. https://halshs.archives-ouvertes.fr/halshs-01132740/document. Accessed 21 April 2020.

STATEC. 2014. Home ownership. https://statistiques.public.lu/catalogue-publications/regards/2014/PDF-26-2014.pdf. Accessed 4 February 2020.

STATEC. 2017. Risk-of-poverty indicators. https://statistiques.public.lu/stat/TableViewer/tableViewHTML.aspx?ReportId=12957&IF_Language=eng&MainTheme=3&RFPath=29. Accessed 4 February 2020.

STATEC. 2019. Luxembourg en chiffre 2018. https://statistiques.public.lu/catalogue-publications/luxembourg-en-chiffres/2018/luxembourg-chiffres.pdf. Accessed 4 February 2020.

Findings and conclusions

Tobias Kuttler and Massimo Moraglio

Abstract

This final chapter offers the main conclusions for both the conceptu-alisation of mobility poverty and its empirical application in the field before discussing a possible roadmap towards alleviating mobility poverty. Applying the concept of mobility poverty to the transport sector should contribute to a fundamental paradigm shift towards achieving mobility justice and equity.

Overview

This final chapter pieces together the results from an extensive analysis of secondary sources and the fieldwork conducted in the context of the Horizon 2020 project "HiReach". The chapter starts with a summary of the main pillars of the mobility poverty concept. Second, the results from the fieldwork will be summarised and discussed, differentiating between planners' and other stakeholders' perspectives on the one hand and the end users' perspective on the other. Here, the opportunities and challenges of alleviating mobility poverty in the study regions are also highlighted, considering the specificities of each study region. Combining these results from two different research methodologies, the conclusion draws the linkages between mobility poverty and social exclusion and proposes a road map and fields of intervention for future action.

Towards a conceptual understanding of mobility poverty

Mobility inequality and injustice are phenomena caused by a number of interrelated processes. As we argued in the introduction to this volume, without acknowledging and analysing the underlying structural disadvantages of mobility injustices and inequalities, policy response will be piecemeal and have limited effect. Therefore, we proposed shifting the conceptualisation of transport poverty to mobility poverty. The main pillars of such adjusted sensitivity in analysing and responding to mobility disadvantage can be summarised as follows:

1   **Mobility and immobility are two sides of the same coin.** A world of seamless mobility is as much characterised by highly mobile societies as by barriers and frictions to mobility that disadvantage those who already experience discrimination (due to gender, race, ethnicity, class, caste, colour, nationality, age, sexuality, disability). Without thorough scrutiny of such underlying conditions, the analysis of mobility poverty remains incomplete.

2   **Mobility poverty needs to be understood in a dynamic relationship with high mobility.** Technological innovation in communication and transport is constantly creating new options for travelling and interaction. Mobility needs potentially grow with the growing availability of options and are interrelated with social, spatial and technological change. This challenges rather static understandings of accessibility and activity participation. The experience of mobility poverty may also occur or be aggravated due to policies and markets focusing on the needs of the highly mobile and the most profitable transport connections.

3   **We must understand mobility needs by differentiating between "mobility" – the actual movement – and "motility" – the potential to move.** Mobility needs and preferences are not just the outcome of an individual's social position and spatial location; mobility needs are also produced and altered according to an individual's biography as well as future aspirations and plans.

    Thus, a person's motility is an indication of one's "mastering" of mobility options, which is closely connected to one's "mastering" of life in general. High motility can express itself in low mobility, but in such cases low mobility will most likely not lead to social disadvantage, mobility poverty or social exclusion. For this reason, it is most important to understand the motility of a person as a kind of capital that is not only determined by an individual's level of access, but also produced by individual skills and competences and shaped by aspirations and attitudes and her/his biography.

4   **Mobility poverty is contextual and relational.** Mobility poverty is experienced differently across European regions. Individual material poverty coupled with an overall poor standard of living, rural/urban deprivation and poor quality of public transport services are strong indicators of an incidence of mobility poverty; however, such a perspective may obscure pockets of mobility poverty in well-developed, advantaged regions.

    Evidence from fieldwork presented in previous chapters suggests that incidents of mobility poverty can have a stronger exclusionary impact on disadvantaged individuals in societies that are highly mobile. Furthermore, in highly mobile societies, hidden and unmet mobility needs may be more difficult to identify. Universal access to public transport can be an important levelling field in economically vibrant regions; hence, the decision of the state of Luxembourg may be a significant step towards creating fair and just access to public transport. Its positive effects on individuals at risk of social exclusion need to be scrutinised thoroughly and proven in further studies.

5   **Mobility poverty often results from a combination of different social disadvantages.** Those most vulnerable to mobility poverty are people experiencing material deprivation linked to physical impairment, migrant or ethnic minority background, single parenthood and different socio-demographic characteristics (being young, old, with issues related to gender). Therefore, the impact of multiple social disadvantages should always be analysed thoroughly when trying to understand experiences of mobility poverty.

6   **We need to understand the differences between mobility-related disadvantage, social disadvantage and social exclusion, but also the linkages.** Mobility-related disadvantage and mobility-related social exclusion are not synonymous with each other: it is possible to be socially excluded but still have good access to mobility options or to be disadvantaged in one's mobility but highly socially included. Social exclusion is determined by many more factors than just mobility; it may even be the least important factor.

7   **Spatial aspects are often only implicitly taken into account in the analysis of mobility-related disadvantage** via the observation of social disadvantage and transportation-related disadvantages. However, the impact of spatial factors on mobility disadvantage is often poorly understood because simplified assumptions are made of the relationship between spatial dynamics and social position. Hence – in addition to a spatial analysis of accessibility at the micro-level – an analysis needs to consider the impact of globalisation on European regions and the differentiated effects on urban and rural regions. While each urban or rural region exhibits particularised development patterns, some far-reaching dynamics can be observed across Europe, such as socio-spatial segregation, urban deprivation, gentrification and re-urbanisation. Peri-urbanisation is recalibrating the historic centre-periphery relationship between cities and rural areas in Europe.

8   **An analysis of mobility poverty should not be limited to viewing the basic mobility needs of everyday life.** Such a perspective misses important aspects that are considered crucial for meaningful life. This may be, e.g. the importance of social interaction and co-presence that are usually connected to leisure trips and therefore beyond everyday life. Proponents of transport equity often argue that basic needs should be met and activity participation should be enabled, both being part of everyday life. They thus argue in favour of "accessibility for all" and not

"mobility for all". While this is easily justifiable (e.g. for environmental reasons), the perspective of the "right to mobility" complicates the matter of transport poverty.

9 **Virtual mobility needs to be analysed together with corporeal mobility.** A person's mobility patterns cannot be understood without understanding one's use of communication tools. The use of communication tools may replace, supplement or create new needs of mobility or conditions of immobility respectively. The interaction between virtual and physical mobility may differ from person to person, with very different outcomes. Furthermore, whether mobility is a choice or compulsion is increasingly difficult to tell.

Fieldwork results

The fieldwork presented in the previous chapter constitutes a pioneering approach towards a better understanding of transport poverty and mobility poverty. Conducting a series of interviews and focus group sessions in six different countries within a large range of social groups and geographical diversity (urban, peri-urban and rural) has improved our ability to recognise and understand mobility-related disadvantages. Since the fieldwork also targeted stakeholders, a more comprehensive outline of the situation in the areas investigated can be offered.

As mentioned, the fieldwork has been developed addressing both end users and stakeholders, considering different social and geographical layers. In the following paragraphs, we will first outline the stakeholders' outcomes, then we will devote our attention to summarising the end users' inputs, and finally we turn towards the opportunities and challenges to alleviating mobility poverty in the study regions.

Stakeholders' voices

The consultation of the stakeholders in the study regions offered some remarkable outcomes. The starting point is a **diversity of understanding of mobility-related disadvantages in the different regions.** On the positive side, advocates of marginalised groups as well as the managers of bottom-up transport initiatives are sensitive to the problems. But, on the other hand, there is a sort of vague and unfocused awareness on the part of more 'classic' public transport suppliers.

There is growing attention of the needs of more vulnerable groups and the necessity to offer more differentiated transport services. Some stakeholders are aware of the diverse social layers' different needs, but we can define **two bottlenecks which impede the implementation of innovative policies.**

First, the mind-set of many stakeholders is still focused on users' physical impediments or low income as the main (if not only) limitations in accessing public transport. This leads to actions towards making transport accessible to anyone with physical impediments or to offering discounted fees for the use of public transport. However, despite the initiatives in the past decades to make public transport accessible to everyone – which do not exist everywhere and are not always successful as the case studies have shown – we still witness an **overarching concept of transport service in which the users are depicted as physically healthy, fully aware of the service and fully able to take advantage of it.**

This leads to a second issue: many transport providers approach their service with a product-driven attitude, **without caring enough about customer needs.** In this mind-set, the customer is an undifferentiated user and the transport supplier takes her/his ability to cope with the service for granted. Furthermore, users' needs are too often portrayed as limited to home-work or home-school commuting, without further investigating any possible additional requirements. Now, considering that public transport is often used by captives, we can understand that this can indeed be a big issue, which leads to a mismatch of demand and supply.

So, overall, while transport managers have some awareness of mobility poverty experienced by many social groups, they still use the binary categories of:

- Users depicted as "normal" and "exceptional"; and
- Services defined as i) "public" and scheduled versus ii) "private" and schedule-free.

The other important issue is the question of budget, which should not be underestimated. While we can say that public transport suppliers do not always target all the end users' needs, it is remarkable to report that they face budget constraints, which hamper the quality of their service. The budget available varies according to areas and countries, but is based on a rather traditional depiction of users more or less everywhere. Trapped in a still predominantly product-driven mind-set, **budget constraints push the management to reduce services and keep a "business-as-usual" attitude**, while they lack knowledge, resources and incentives to pursue innovation.

Considering the disruptive changes on transport markets and the peculiar difficulties of some social groups we face a dilemma:

On one hand, the lack of supply by traditional transport operators leaves us with **plenty of opportunities to develop innovative projects**; but, on the other hand, too often such new transport solutions (ride-hailing, flexible transport, car-sharing, bike-sharing) address the needs of **"strong" users, those with digital skills, great cognitive abilities and, last but not least, a credit card.**

End users' voices

Among the most interesting outcomes of the focus group sessions with the end users, we should first mention **that those engaged in the discussion were**

very articulate and communicative. We are fully aware of the (inherent and unavoidable) limitations of focus groups in the sense that they give a louder voice to those who are already vocal. Still, there was great interest by users to discuss the topic.

The second element to mention here is the **wide range of options presented in order to combat mobility-related disadvantages**. This goes from very basic requests, such as better footpaths and safe cycle parking (as in the case of Buzău) to suggestions for bottom-up and peer-to-peer car-sharing (as for Naxos and Small Cyclades) and tailor-made, flexible, on-demand services (as demanded in Guarda).

As a third observation, there is often (but not everywhere) **a lack of trust towards public authorities and more specifically towards public transport suppliers**. This is sometimes the consequence of poor services and sometimes the result of users' own high expectations. It is also important to notice that this is often accompanied by a sort of fatalism, which impedes any action and leaves users waiting for top-down actions.

Many users are trapped by a total dependence on cars, which are depicted as a mixed blessing. On one hand, for those who can drive a car (or travel in it as a passenger), private motor vehicles are the only reliable modes of transport at the end of the day. In personal situations of low income, this car dependency, without realistic alternatives, makes low-income groups highly vulnerable to policies that seek to limit car use (pricing, taxation or a ban on highly polluting old vehicles).

On the other hand, in the focus group sessions, it became clear that men usually have priority in the use of automobiles, which leaves **women with fewer opportunities**, those being very challenging and time-consuming. Worse than this, a still dominant and aggressive use of cars is also reported. Besides the related risks, this limits any opportunity to share roads and ultimately this hampers the development of other forms of transport, such as cycling.

In a more theoretical stance, we should also note that mobility poverty is the product of concomitant elements. While in academic debate there tends to be a focus on singular aspects, such as language or physical barriers, the focus group sessions revealed that **we should rather consider mobility poverty as a multi-layered phenomenon**. Indeed, while the categorisation of social and spatial layers is important from an analytical perspective, the end users confirmed that everyone, in practice, belongs to more than one group.

This overlapping accentuates and increases the risk of mobility poverty. The focus groups also highlighted many of the assumptions that were made based on earlier studies. For instance, the cases of Naxos and Iraklia magnify the traditional mobility problem of remote areas, adding island isolation to the generally rural difficult accessibility. The case of Naxos and Iraklia also clearly showed that children and the elderly are those who pay the highest price: we have clear evidence of geographical isolation and poor transport systems further triggering social exclusion. Also, the relation between mobility poverty and geographical scale is evident, again comparing Naxos (18,904 inhabitants) with Iraklia (141).

The **cognitive appropriation and understanding of mobility options** was also addressed in the focus groups. In Romania, children and young people are fully aware of the bicycle's socio-technical system, asking for it to be improved (bike lanes, facilities to park bikes securely) in order to be able to go to school by bike and thus reduce their dependency on other modes of transport. Conversely, in Germany, senior drivers, especially males, find it difficult to change from car use to buses, declaring they find it difficult to understand how public transport works.

This leads to another observation: not only in Germany, the "younger" elderly (also when retired) have very active lifestyles. It is an important outcome, which needs further analysis (and also to be leveraged for bottom-up initiatives) and to avoid stereotypical images of this social group.

Opportunities and challenges to alleviating mobility poverty in the study regions

Some user needs target the **very basics of the urban structure**: the request by school facilities for safe road crossings and cycle parking (e.g. for young Romanian people) and the need for well-maintained footpaths (for blind people) are indeed related to elementary infrastructure that can be realised with a very low budget and low investment.

The issue of safety, both real and perceived, as mentioned in other focus groups, is also relevant and often beyond the control of any transport operators. Still, addressing these concerns can make the difference and unleash great potential.

However, we can also list **simple requests to public transport operators**, for example to provide more selling points where people can buy a ticket. Possibly, the operator is aiming to reduce distribution and retailing costs, but the scarcity of sales channels also becomes a burden for passengers, especially those who cannot afford monthly subscriptions. Digitisation of the information is also requested. Taking action to meet these requests, which are definitely low profile, can indeed increase the quality and accessibility of existing services, thus enhancing their appeal.

Once we aim to **define innovative transport regimes coping with mobility poverty, we face some challenges and some opportunities**. While we have an array of inspiring grassroots initiatives (such as informal car-pooling and peer-to-peer car-sharing) at our disposal, we also encounter a lack of trust towards public authorities and the very poor reputation of existing public transport services.

The lack of trust towards local authorities is evident and it triggers a self-fulfilling prophecy: the service is perceived as poor and only for "captives" so the suppliers have no incentives to improve the service, which causes them to become even less appealing, and so on. A better understanding of

Table 19.1 Opportunities and challenges to alleviating mobility poverty in the study regions

Country	Area	Opportunity			Challenge			
		Bottom-up	*Positive view of biking and other "poor" systems*	*Openness to shared transport systems*	*Weak knowledge by suppliers and planners*	*Top-down dominant approach*	*Weak trust in PT and authorities*	*Car as dominant mode*
Germany	District of Esslingen	✓						✓
Greece	Naxos and Small Cyclades	✓	✓	✓	✓		✓	✓
Italy	Inner Area Southern Salento				✓		✓	✓
Luxembourg	North and south-east Luxembourg			✓				✓
Portugal	Guarda			✓	✓	✓	✓	✓
Romania	Buzău		✓	✓	✓	✓	✓	✓

Source: Authors

user needs and improved actions by policy makers to address mobility and mobility poverty are necessary.

Moving back to distrust, this is an important point in launching a new service, which should rely on the support of local service providers, but also avoid negative labelling regardless of its quality.

On the other hand, there is a general affirmative understanding and use of "alternative" mobility options and such a positive attitude should be capitalised by new initiatives. This can also be said for shared transport systems, especially in rural regions.

Conclusions

From the spatial and social analysis presented above, conclusions can be drawn for three main aspects that are crucial to alleviate mobility poverty:

1 Mobility poverty and the risk of social exclusion;
2 Approaches to alleviate mobility poverty; and
3 Fields of intervention.

Mobility and the risk of social exclusion

It was shown in the analysis that social disadvantage in conjunction with mobility-related disadvantage leads to mobility poverty. However, as already emphasised previously, mobility poverty does not necessarily lead to social exclusion.

The analysis revealed the circumstances under which a high risk of social exclusion due to mobility poverty may arise. When linking the conclusions from the social and spatial analysis together, it is revealed that experiences of mobility poverty are a combined outcome of social disadvantage, negative spatial conditions and unmet mobility needs.

This cross-sectional observation reveals that the **risk of social exclusion due to mobility poverty is highest when two or more of the following conditions interact** (see Figures 19.1 and 19.2):

- **Social aspects:**
- **Experience of multiple social disadvantages, especially when low income levels and unemployment are involved:** The conjunction of different social disadvantages and vulnerabilities increases the risk of social exclusion. Incidences that frequently appear are, for example, old age in conjunction with mobility impairment or old age and living in remote rural areas. Other examples that were shown are disabled young people and migrant women. In all cases, low income, unemployment and precarious working conditions substantially increase the risk of social exclusion due to mobility poverty (see Figure 19.1).
- **No car ownership or forced car ownership:** the risk of social exclusion is higher when vulnerable individuals do not have access to cars. Such a risk is particularly prevalent in rural areas, where public transport availability is lower, income levels are lower and distances to opportunities are higher than in urban and peri-urban areas. Car ownership is

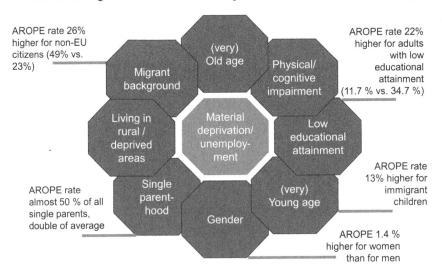

Figure 19.1 Impact of material deprivation on mobility poverty.
Source: Authors, with data from Eurostat

almost unavoidable in such areas, which poses a high cost burden on materially deprived individuals. The money that is spent for mobility is then missing in other essential areas of life.

- **Spatial aspects:**

  - **Low accessibility level**: the risk of social exclusion increases substantially for those individuals whose access to mobility options and access to opportunities is low. This is the case for remote rural areas throughout Europe, but particularly in eastern and southern Members States of the EU. Also, peripheral urban locations can have low accessibility levels. However, there is no determinism between urban peripherality and inaccessibility.
  - **Economically declining area**: economically declining regions can be found all over Europe and all three spatial levels (urban, peri-urban and rural) are affected. When economic decline leads to the

Figure 19.2 High risk of social exclusion due to mobility poverty.
Source: Authors.

outmigration of the young and skilled population, coupled with decaying infrastructure and diminishing service levels, experiences of mobility poverty in such areas substantially increase the risk of social exclusion.

- **Low mobility/motility level**: Mobility is the primary form of (social) capital in advanced societies and crucial for sustaining social networks. The necessity of being mobile can be a serious burden for vulnerable social groups. Thus, unmet mobility needs and low mobility levels can lead to relative disadvantages vis-à-vis those being highly mobile. However, as has been shown, even more important for freedom of choice is the ability to decide when, how and where to move or to stay put. In certain situations, the decision to remain immobile will benefit an individual more than the decision to be mobile.

Approaches to alleviate mobility poverty

The state-of-the-art definitions of transport poverty (Lucas 2012; Lucas et al. 2016) understand transport poverty as the combination of an experience of social disadvantage and transport-related disadvantage. Transport poverty can lead to social exclusion, which reinforces both transport disadvantages and social disadvantages. Whether an individual is transport poor or not is determined by (at least) five conditions: (i) availability and accessibility of transport, (ii) locations and opportunities; (iii) affordability of transport; (iv) available time budget; and (v) adequacy of travel options. The occurrence of one single condition can lead to an individual experiencing transport poverty.

Hence, the mobility needs for each vulnerable group need to be analysed and, accordingly, these basic transport conditions need to be improved to create inclusive mobility options for vulnerable individuals.

As shown, depending on the needs of different social groups, some conditions are more important than others. In terms of adequacy, for elderly people and women, safety in transport is a paramount precondition for using public transport options. Negative experiences can lead to the avoidance of public transport. In addition to safety, healthy travel conditions are crucial for children and young people. Availability (including reliability), accessibility and affordability are crucial for those on low income and with no access to cars.

The analysis of social and spatial disadvantages supports a focus on increasing accessibility for all vulnerable groups in order to increase the potential for participating in activities.

Following the rationale of this volume, the shift from transport poverty to mobility poverty requires recognising additional factors of mobility disadvantage. One of these factors is that low mobility individuals can experience relative disadvantages in highly mobile societies. As previously shown, individuals with low levels of mobility may have unmet or unrecognised mobility needs that are out of reach for these individuals, due to lifelong experiences of disadvantage, habits and routines or gender roles.

Hence, increasing accessibility can secure basic needs, but life satisfaction and mental well-being may still be reduced due to the inability to "keep up" with others in society.

Thus, **in addition to accessibility, it is crucial to increase motility – the potential to move – for members of vulnerable social groups**. Here, it is important to remember Schwedes et al. (2018) who highlighted that mobility comprises also mental flexibility and agility. It is important for members of vulnerable social groups to increase their mental horizon and have the capacity to plan and shape their own lives. Only then will the spaces of opportunity for disadvantaged individuals become larger (Figure 19.3).

Due to the significance of early travel socialisation as well as the importance of travel for the accumulation of social and network capital at an early age, disadvantaged children and young people should have the opportunity to travel and experience a wide range of mobility solutions. Also, for elderly people, not only is access to basic services crucial, but also the ability to move is paramount to being part of social networks and maintaining a meaningful life in old age.

Hence, elderly and mobility-impaired people need to be informed and enabled to explore all the different mobility options available. While traditional gender roles and models that characterised women's mobility are steadily becoming less common, it is important to challenge mobility policy and planning by including gender perspectives more strongly in these domains.

A comprehensive approach to alleviating mobility poverty should therefore tackle the underlying, structural social disadvantages. This means that formulating policy and planning needs to intervene in policy sectors that are upstream of transport policy. With Sheller (2018) it can be argued that four different forms of justice need to be achieved before transport policy can be

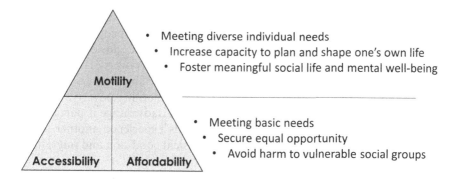

Figure 19.3 Approaches to alleviating mobility poverty.
Source: Authors.

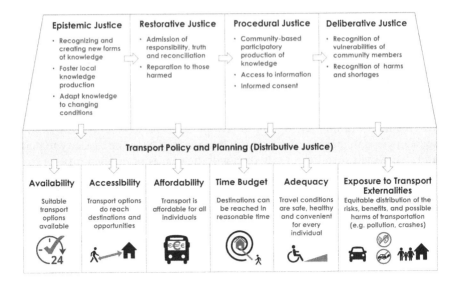

Figure 19.4 Road map to alleviating mobility poverty.
Source: Authors, based on Sheller 2018 and Lucas et al. 2016.

made effective. Linking mobility justice and transport policy may result in a comprehensive and concrete road map to alleviate mobility poverty (see Figure 19.4).

Fields of intervention

This volume highlights incidences of mobility poverty that suggest certain fields of urgent intervention in order to prevent the social exclusion of vulnerable parts of the population:

- **Focus on people at risk of poverty:** the share of people at risk of poverty in Europe is substantial: in 2015, almost 119 million people, or 23.7% of the population, were at risk of poverty or social exclusion in the EU-28. As shown above, the experience of material poverty is often associated with material deprivation. It must be assumed that a large part of those at risk of poverty are also at risk of mobility poverty. The risk of social exclusion due to mobility-related disadvantage is particularly high when materially deprived individuals experience another social disadvantage related to age, gender, physical condition and migrant or minority status.
- **Focus on women:** in this volume, it has been acknowledged that women experience substantial disadvantages in their mobility due to a variety of factors such as lower incomes, gender roles and access to modes of

transport. Furthermore, they are more likely to be at risk of poverty and social exclusion. As the ageing of European societies continues, elderly women will represent a substantial part of the future population of the EU.

- **Focus on children:** children and young people suffer the most from inadequate mobility options. If inadequate transport services result in barriers to education, training and employment at a young age, they will experience substantial repercussions as they grow older.
- **Focus on deprived and peripheral urban areas as well as peri-urban areas:** more and more people are living in metropolitan areas due to the availability of jobs. However, many people are pushed out of cities – due to inadequate and expensive housing – into peripheral urban areas or peri-urban areas well beyond the city limits. Others remain in deprived inner-city areas. These types of areas may experience inadequate public transport coverage or car dependency that contributes to the marginalisation of vulnerable individuals.
- **Focus on economically declining regions and remote rural areas:** the population in such regions is ageing and becoming smaller. The attention of policy and planning is increasingly directed at metropolitan regions where the majority of the EU population lives. Hence, it is important to continue the strategic development of instruments for old industrial and remote rural areas that tackle the further decline of these regions. In order to secure adequate standards of living and potentially attract new economic activities, mobility-related interventions are among the many interventions needed in these areas.

References

Lucas, Karen. 2012. "Transport and social exclusion: Where are we now?" In *Transport Policy* 20: 105–113. https://doi.org/10.1016/j.tranpol.2012.01.013.

Lucas, Karen, Giulio Mattioli, Ersilia Verlinghieri, and Alvaro Guzman. 2016. "Transport poverty and its adverse social consequences." In *Proceedings of the Institution of Civil Engineers* 169 (TR6): 353–365. https://doi.org/10.1680/jtran.15.00073.

Schwedes, Oliver, Stephan Daubitz, Alexander Rammert, Benjamin Sternkopf, and Maximilian Hoor. 2018. *Kleiner Begriffskanon der Mobilitätsforschung*. 2nd edition. IVP-Discussion Paper. Berlin. https://www.econstor.eu/bitstream/10419/200083/1/ivp-dp-2018-1.pdf. Accessed 21 April 2020.

Sheller, Mimi. 2018. *Mobility Justice: The Politics of Movement in an Age of Extremes*. Brooklyn: Verso Books.

Index

Printed in the United States
By Bookmasters